A Short Introduction to Geospatial Intelligence

A Short Introduction to Geospatial Intelligence explains the newest form of intelligence used by governments, commercial organizations, and individuals. Geospatial intelligence combines late 20th century historically derived ways of thinking and early 21st century technologies of GIS, GPS, digital imaging satellites and communications satellites to identify, measure, and analyze the current risk in the world.

These ways of thinking have developed from military engineering, cartography, photointerpretation, and imagery analysis. While the oldest example dates back to the early 16th century, all the ways of spatial thinking share the common thread of being developed and refined during conflicts to help military leaders make informed decisions prior to action. In the 21st century—thanks in great part to advances in digital precision technology, miniaturization, and the commercialization of satellites—these ways of thinking have expanded from the military into various other industries and sectors including energy, agriculture, environment, law enforcement, global risk assessment, and climate monitoring.

Features:

- Analyzes human and algorithmic models for dealing with the challenge of analytic attention, in an age of geospatial data overload
- Establishes an original model—envisioning, discovery, recording, comprehending, and tracking—for the spatial thinking that underpins the practice and growth of this emerging discipline
- Addresses the effects of small satellites on the collection and analysis of geospatial intelligence

A Short Introduction to Geospatial Intelligence describes the development of the five steps in geospatial thinking—envisioning, discovery, recording, comprehending, and tracking—in addition to addressing the challenges, and future applications, of this newest intelligence discipline.

Jack (John) O'Connor has directed the Master of Science in Geospatial Intelligence at Johns Hopkins University since 2018. For over 30 years, in the Central Intelligence Agency (CIA) and Department of Defense (DoD) organizations, he led imagery and geospatial analysis in combat support, national intelligence, diplomatic initiatives, and disaster support, as well as regional, environmental, economic, and social analysis. He has received numerous government awards, including the Intelligence Community's Galileo Award and the National Intelligence Medal of Achievement. O'Connor holds a Master of Arts from Bryn Mawr College, an Bachelor of Arts from St. Joseph's University, and an Associate of Arts from Delaware County Community College. He is a member of the US Geospatial-Intelligence Foundation and the Medmenham Association.

A Short Introduction to Geospatial Intelligence

Jack (John) O'Connor

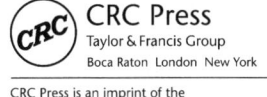

CRC Press
Taylor & Francis Group
Boca Raton London New York

CRC Press is an imprint of the
Taylor & Francis Group, an **informa** business

Designed cover image: CIA/PIC

First edition published 2024
by CRC Press
2385 NW Executive Center Drive, Suite 320, Boca Raton FL 33431

and by CRC Press
4 Park Square, Milton Park, Abingdon, Oxon, OX14 4RN

CRC Press is an imprint of Taylor & Francis Group, LLC

© 2024 Jack (John) O'Connor

ISBN: 9781032566993 (hbk)
ISBN: 9781032566948 (pbk)
ISBN: 9781003436836 (ebk)

DOI: 10.4324/9781003436836

Typeset in Sabon
by Newgen Publishing UK

For Kathy

Contents

Preface

This short book grew out of an attempt to answer a question asked by some of the academic members of the United States Geospatial Intelligence Foundation. The question was: What makes GEOINT, GEOINT? That question would lead to a 2019 discussion about the definition and meaning of geospatial intelligence, and a virtual presentation in February 2020. That presentation, in which I participated with Dr. Todd S. Bacastow of Penn State University, Dr. Robert T. Clark of Johns Hopkins University, Dr. Camelia Kantor then of the USGIF, and Dr. Tim Walton of James Madison University, and the encouragement of my colleagues, led me to capture what follows.

All statements of fact, opinion, or analysis expressed are those of the author and do not reflect the official position or view of the CIA, or any other US government agency. Nothing in the contents should be construed as either asserting or implying US government authentication of information or Agency endorsement of the author's views. This material has been reviewed by the appropriate intelligence agencies to prevent the disclosure of classified information.

Illustrations

Acknowledgments

My thinking and writing on this subject have been improved by the comments and questions so graciously provided by Dr. Renny Babiarz, Dr. Todd S. Bacastow, Joseph Caddell, Jr., Dr. Robert M. Clark, Jim David, Tim Holden, Dr. Camelia Kantor, Thom Kaye, Veli-Pekka Kivimaki, Mark G. Marshall, Kevin Pomfret, Joshua Rooksby, Dr. Mark Stout, Dr. Peter S. Usowski, Patrick Widlake, and Dr. Tim Walton.

For permission to use the illustrations in this book I would like to thank Robbin Brooks of the National Geospatial-Intelligence Agency, Dr. Lauren Gardner of the Center for Systems Science and Engineering at Johns Hopkins University, Stephanie Kurasz of the Smithsonian Institution's National Museum of American History, Daniel Partridge of the Royal Collection Trust, and Ruth Pulley of the Medmenham Association.

My thanks also go to the anonymous manuscript readers at CRC Press. Additionally, I would like to express my appreciation for my students and colleagues in the Advanced Academic Programs in the Krieger School of Arts and Sciences at Johns Hopkins University, along with the librarians in the Fairfax County Public Library and the Johns Hopkins Library. All the above deserve credit for improving this book; I alone am responsible for any errors.

Introduction

Five intellectual activities comprise geospatial intelligence: envisioning, discovery, recording, comprehending, and tracking. While these activities have existed as long as men and women have looked at and thought about the planet, geospatial intelligence developed in the early 21st century. It came into being when newly developed technologies invented in the late 20th and early 21st centuries—the Global Positioning System (GPS), Geographic Information Systems (GIS), digital imaging satellites, commercial small satellites, and the global internet—transformed the human ability to capture changes on the planet and to the planet.

This book is written for readers curious about the long human tradition of using these intellectual activities, and the recent advances in digital communications on earth and in space, to make geospatial intelligence possible. This book does not presume a knowledge of the history, technology, mathematics, or science that informs this new discipline. It outlines these intellectual activities and the technologies that have enabled the discoveries, comprehension, and tracking that this unique discipline and its forerunners have brought about over the past hundred years. For the readers interested in the historic, mathematical, technological, or scientific background, the footnotes and the suggested readings provide directions for future reading or viewing.

This book relates Human Intelligence to one type of intelligence customarily used by nations, businesses, and other organizations. All governments perform intelligence work to help reduce future risks and avoid strategic surprises. Traditionally, intelligence helps governments and organizations make better decisions and take less risky actions. It involves the analysis of information, frequently partial, ambiguous, or deceptive information gathered from people (HUMINT), collected from intercepted signals and communications (SIGINT and COMINT), gathered from open or unclassified sources in the press or media (OSINT), or collected from measurements and sensors, like radioactivity (MASINT). Governments and organizations expend all this effort to enable the recipients of intelligence, usually policymakers or military decision-makers, to think or act differently and to assess risk better.

Geospatial intelligence (GEOINT) differs linguistically from all the other forms of intelligence in that its shortened form does not relate to the source of the information used to detect changes, as all the other "ints" do. To explain the intellectual activities or Human Intelligence used to create geospatial intelligence, Human Intelligence will be capitalized throughout and geospatial intelligence will be written in lowercase.

Most of the stories and examples in this book are military, and describe survival and destruction. These two human motivations—personal survival and destruction of others—propelled the technological developments and developed the intellectual activities—envisioning, discovery, recording, comprehending, and tracking—that this book describes. The history of human conflict has displayed a mortal serious-ness in using technology to enable a better means of discovering and killing whatever group has been defined as an opposing force. The non-military stories of geospatial intelligence treat human efforts to deal with other kinds of threats. Military history encompasses terrain, discovery, concealment, camouflage, and tracking techniques. Many of these stories relate the importance of finding a location. For explorers and navigators, discovering their own location was essential for survival and returning to their starting point. For military leaders, discovering the location of others, particularly the enemy, was equally important for an advantage in deciding how to destroy the enemy, emerge victorious in conflict, and sur-vive. These two human motivations—personal survival and destruction of others—propelled the technological developments and impelled the intellectual activities—envisioning, discovery, recording, comprehending, and tracking—described in this book.

Geospatial thinking and geospatial technologies make up geospa-tial intelligence. The advances in thinking about envisioning, discovery, recording, comprehending, and tracking, as well as the technologies that have enabled and improved location, resolution (the ability of images to detect smaller objects), remote sensing, digital communication, com-putational mathematics, and geography, have coalesced over the past 60 years to enable this 21st-century discipline to emerge. The legal and ethical ramifications of this new intellectual activity and the new tech-nologies have already shaped and will continue to shape the 21st century. In recent years, the developments in geospatial intelligence have begun to shape other aspects of human society, such as energy, the environment, climate change, and trade. But the origin of the term, GEOINT, for geo-spatial intelligence, originated in a bureaucratic budget struggle in the US Defense Department near the end of the 20th century

THE CHALLENGE OF DEFINING GEOSPATIAL INTELLIGENCE

The origin of the term "geospatial intelligence" remains uncertain, but the origin of the shortened form, GEOINT, is known. It was first spoken in a

National Imagery and Mapping Agency (NIMA) offsite meeting in 2001 or 2002 by Rob Zitz, then a NIMA senior manager.[1] NIMA had been created in 1996, as an outgrowth of post-Cold War budget reductions in the United States Department of Defense and Intelligence Community and computer technology developments in the private sector. The NIMA charter stated:

Sec. 442. Missions

(a) National Security Missions.--(1) The National Imagery and Mapping Agency shall, in support of the national security objectives of the United States, provide the following:
  (A) Imagery.
  (B) Imagery intelligence.
  (C) Geospatial information.[2]

The NIMA definition of geospatial information in 1996 was maps, imagery and text, books, and geodetic products. As an agency, NIMA existed from 1996 until 2003, when its third director, James Clapper, changed the agency name to the National Geospatial-Intelligence Agency. The act that renamed the agency provided the first definition of geospatial intelligence in law:

The term "geospatial intelligence" means the exploitation and analysis of imagery and geospatial information to describe, assess, and visually depict physical features and geographically referenced activities on or about the earth. Geospatial intelligence consists of imagery, imagery intelligence, and geospatial information.[3]

The NGA definition of geospatial intelligence in law did not change in content from the NIMA charter, but in practice geospatial intelligence began to stretch the legal definition. This developing definition has been noted by academics and practitioners since the term was legally defined. The discomfort with the definition of geospatial intelligence can be considered a subset of the general discomfort that practitioners and academics have with the definition of intelligence.[4] In 2009 Todd S. Bacastow and Dennis Bellefiore made the first attempt to align the definition of geospatial intelligence with a definition of intelligence:

Geospatial Intelligence is actionable knowledge, a process, and a profession. It is the ability to describe, understand, and interpret so as to anticipate the human impact of an event or action within a spatiotemporal environment. It is also the ability to identify, collect, store, and manipulate data to create geospatial knowledge through critical thinking, geospatial reasoning, and analytical techniques. Finally, it is the ability to ethically collect, develop, and present knowledge in a way that is appropriate to the decision-making environment.[5]

Bacastow and Bellefiore suggested an alternate acronym—GeoIntel—to distinguish the intellectual discipline involved in creating this kind of intelligence from the product and the process. In 2016, John Oswald and Scott Simmons, in an article in the *Geospatial Intelligence Review*,[6] developed the idea of distinguishing the mental activity involved in the creation of geospatial intelligence from the mostly classified source material—airborne and satellite imagery and geospatial databases. Their article distinguished geospatial analysis from geospatial intelligence.

In their chronicle of the beginnings of geospatial analysis in the NIMA, Oswald and Simmons defined the US government's role:

> Academe and industry had been combining GIS and remote sensing to address commercial activity, agriculture, emergency services, and urban planning. Government imagery and geography professionals also recognized the potential benefits. Inside NIMA, senior managers like Robert E. Zitz, Irvin Buck, and Roberta E. (Bobbi) Lenczowski set geospatial analysis in motion—but at a faster pace and with greater complexity than in the private sector. NIMA had the power of sources, technologies, and tradecraft found only in the Intelligence Community (IC) and Department of Defense.[7]

While Oswald and Simmons noted how the defense and intelligence sectors of the US government brought intelligence sources, technologies, and tradecraft to the GIS capabilities of the private sector, their article did not discuss the ends or the new kinds of analysis that analysts used this technology to achieve.

In the same year, Robert M. Clark and Darryl Murdoch proposed a new definition of geospatial intelligence that bridged the difference between the government and commercial sectors and applications. Clark and Murdoch, building on the Bacastow and Bellafiore definition, stated:

> a more comprehensive definition of GEOINT might be this: GEOINT is the professional practice of integrating and interpreting all forms of geospatial data to create historical and anticipatory products used for planning or that answer questions posed by the decision-makers.[8]

Murdoch and Clark's definition moves away from the technological inputs and focuses on the analytic activities of geospatial analysts. More recently in 2020, Clark developed his own definition:

GEOINT requires analyzing a situation and creating a product that

- is anticipatory—it deals with the future;
- involves some type of human activity;

- draws on knowledge of the Earth (its surface, whether on land or sea, and the objects on or beneath that surface); and
- provides an information advantage to someone who must make a decision.[9]

Clark's definition encompasses more of the human role in creating geospatial intelligence. This book acknowledges the previous definitions and builds on Clark's definition to identify and explain the five characteristics of geospatial intelligence. They are in order

Envisioning or the human combining of imagination and measurement to see changes on the planet differently than they have been seen before. Initially, envisioning involved human vision, but with the evolution of sensing and digital technology, it now encompasses using sensors to track data about changes that are not visible to the human eye, by using different parts of the electromagnetic spectrum.

Discovery follows envisioning. It also can have its basis in human vision or computer sensors, but it is a new awareness of change or activity on the planet, frequently, but not always man-made activity. Some discovery results from diligent efforts over time, but chance plays a part in some discovery.

Recording follows discovery, as it is necessary to create a record of the changes. Initially, these records were analog drawings, records of measurements, and textual descriptions, but the invention of chemically based technologies allowed for photography, and later digital computer-based technologies, such as GPS and GIS, enabled precision records that are much easier to update for preservation and future use.

Retrospective review of the recording of observations enables *comprehending*, or the ability to put physical and human changes into a functional or regional context, to track the chronology of events, and to assign meaning to the observations that explains their causes, effects, or processes.

And finally, when the frequency of observation is done in real time, *tracking* is possible so that the information is as current as possible and the discovery, comprehension, and recording can be accelerated to keep up with a dynamic pace of change. Tracking, with the modern technologies of digital video cameras, armed unmanned aircraft, and satellite communications, can change rapidly into military targeting and more rapidly into military operations.

The review of the existing definitions of geospatial intelligence illustrates the possibilities now enabled by digital, spatial, orbital, and terrestrial technologies. Through GPS, precision locations are now commonplace, and through GIS, humans can revise, record, and analyze spatial information far more rapidly. But new and faster technology explains only part of what enables geospatial intelligence. This book contends that five intelligence activities underpin geospatial intelligence. These activities have existed through the history of cartography and aerial intelligence, particularly photographic and imagery-derived intelligence. And they are comprehensible without a deep grounding in the history of either of these disciplines. But at the core of the historical practices of creating cartography and intelligence is seeing differently.

NOTES

1 Author's conversation with Peter Usowski, 2021, who attended and witnessed the event, and with Rob Zitz, 25 April 2022, who confirmed that the NIMA director at that time, General James King, chose not to embrace either the concept or the term. Robert M. Clark, in his *Geospatial Intelligence: Origins and Evolution*, has one of the origin narratives.
2 www.intelligence.senate.gov/laws/national-imagery-and-mapping-age ncy-act-1996, page 110 stat 2678.
3 www.law.cornell.edu/uscode/text/10/467.
4 The growth of articles and books on this topic can be seen by looking at Michael Warner's two articles on this topic. His first article, "Wanted: A Definition of 'Intelligence,'" in *Studies in Intelligence*, Vol. 46, No. 3 (2002), cites practitioners and government organizations reviewing Intelligence organizations. His second article, co-written with Mark Stout, "Intelligence is as intelligence does," *Intelligence and National Security*, 2018, Vol. 33, No.4, pp. 517–526, adds another category of citations from the academic community, mostly historians of intelligence. Both articles note the multiple definitions and the lack of agreement among academics and practitioners.
5 Bacastow, Todd S. and Dennis Bellafiore. "Redefining Geospatial Intelligence," *American Intelligence Journal*, Fall 2009, Vol. 27, No. 1, p. 40.
6 Oswald, John A. and Scott Simmons. "Geospatial Analysis: Origin and Development in the National Geospatial-Intelligence Agency," *Geospatial Intelligence Review*, 14, No. 1 (2016).
7 Oswald and Simmons, p. 70.
8 Murdoch, Darryl and Robert M. Clark, "Geospatial Intelligence," pp. 114, in *The Five Disciplines of Intelligence Collection,* eds. Mark M. Lowenthal and Robert M. Clark. (Thousand Oaks, CA: CQ Press, 2016).
9 Clark, Robert M. *Geospatial Intelligence: Origins and Evolution.* (Washington, D.C.: Georgetown University Press, 2020), p. 8.

BIBLIOGRAPHY

Bacastow, Todd S. and Dennis Bellafiore. "Redefining Geospatial Intelligence," *American Intelligence Journal*, Fall 2009, Vol. 27, No. 1. Intelligence Support to the Warfighter, pp. 38–40.

Clark, Robert M. *Geospatial Intelligence: Origins and Evolution* (Washington, D.C.: Georgetown University Press, 2020).

www.intelligence.senate.gov/laws/national-imagery-and-mapping-agency-act-1996,page 110,stat2678.

Murdock, Darryl, and Robert M. Clark. "Geospatial Intelligence," in *The Five Disciplines of Intelligence Collection*. Eds. Mark M. Lowenthal and Robert M. Clark, pp. 111–158 (Thousand Oaks, CA: CQ Press, 2016).

Oswald, John and Scott Simmons. "Geospatial Analysis: Origins and Development in the National Geospatial-Intelligence Agency," *Geospatial Intelligence Review*, Vol. 14, No. 1, 2016.

CHAPTER 1

The Genome of GEOINT

In biology, a genome is defined as the whole of an organism's hereditary information encoded in its DNA. The Jam Session, a joint government and civilian analytic experiment to determine what could be learned from aerial photography about Soviet Strategic missiles in November 1957—with its combination of imagery, imagery intelligence, and geospatial information—can be considered the genome of geospatial intelligence. The events leading to this unique experiment illustrate the intellectual activities that constitute geospatial intelligence.

ENVISIONING BEHIND THE IRON CURTAIN

After World War II, the Soviet Union emerged as the great rival of the United States (US) and other Western democracies. The Soviet Union, a closed society, was a very hard intelligence target and reliable information about the vast Soviet interior was virtually impossible to obtain. Its ideology seemed threatening as did its massive military. After the Soviet Union tested its first atomic bomb in 1949, concerns about the Soviet threat grew all the greater. The US government clamored for intelligence to help it understand and counter the threat, especially the Soviet strategic nuclear forces.

The US knew very little about the Soviet Union. It knew less about its military and strategic infrastructure. It knew that the USSR had possessed atom and hydrogen weapons since 1949 and 1953, respectively, and it had no warning before these atomic weapons were initially tested. At the 1954 May Day parade in Moscow, the US learned from the public display that the Soviet Union had built at least a few long-distance jet bombers, but it had no idea how many had been built, where they were deployed, or their capabilities. The surprise discovery at the parade meant that the Soviet Union had at least one way to deliver a nuclear first strike.[1]

After the Soviet Union successfully tested its hydrogen bomb in 1953, President Eisenhower convened an advisory group to help the US learn about the military, strategic, and scientific capabilities of the Soviet

DOI: 10.4324/9781003436836-2

Union.[2] One of the five sections of the panel's report, Project 3, addressed how US intelligence could learn more about the USSR's capability for a surprise attack on the US. Project 3 was chaired by Edwin H. (Din) Land, Chairman of the Polaroid Corporation.

Din Land made his scientific reputation and fortune by envisioning light differently. Extraordinarily intelligent and focused from an early age, Land began studying the polarity of light and began teaching himself from college textbooks on optics and physics while he was in high school. He enrolled at Harvard but stopped attending classes as he could teach himself more and more quickly by reading in the library and experimenting in the lab. He left the Harvard campus and went to the New York Public Library to read and research. Without access to a laboratory, he snuck into Columbia University to perform an experiment that proved that he could create a film that would polarize light rays. He returned to the Harvard as the only undergraduate to teach a graduate seminar in physics and to have been granted his own laboratory. While he never took an undergraduate degree, he formed a company with a Harvard professor, earned several patents related to how to polarize light, and went on to found the Polaroid Corporation.

Land had used his creative talents to support the US military in World War II. After the war, his reputation was such that he was invited by the US Air Force to participate in early studies about how to defend the US mainland from nuclear attack. In the 1952 Beacon Hill Study,[3] in which Land wrote Chapter 7, he outlined the technical and organizational requirements for the reconnaissance aircraft that would become the U-2. But the Air Force was not interested and did not act on Land's ideas.

Two years later Land was invited to participate in President Eisenhower's Technical Capabilities Panel and Land was put in charge of Project 3, the committee on intelligence. Land was a pragmatic genius. He had a rule that a committee could succeed only if it could fit in one taxicab. The Project 3 committee of the Technological Capabilities Panel had only five members. One of its members, James Baker, was the first physicist who used a computer to design lenses for high-altitude photography, and he would design the cameras for the U-2 aircraft. Land also recognized, after his Beacon Hill experience, that he would have to work around the Pentagon bureaucracy.[4]

Land wrote the committee report and sent a memo to Allen Dulles, then Director of the Central Intelligence Agency (CIA). Land had surveyed all the Department of Defense (DoD) research on missile and space development, as well as the work done by think tanks like the RAND corporation. From his research, he recommended a design for a reconnaissance aircraft that became the U-2. Also, he advocated that CIA manage the project, and he combined the best sensor and the best platform to address the intelligence issue that his project had been studying. But Killian and

Land did not include all their findings in the report. Both men briefed President Eisenhower about their recommendations and the challenge of the Pentagon bureaucracy.

But Land was also a realistic pragmatist. In his November 1954 Memo to Allen Dulles, Land recommended a Lockheed design for technical reasons, and also noted the need to accelerate the project as, even if successful, it would have only a few years before the likelihood of detection and shootdown would be likely.[5] Land's work, accomplished between July and November 1954, set in motion the U-2 program at CIA. The U-2 aircraft, designed and created by "Kelly" Johnson at Lockheed with cameras created by James Baker, and the exploitation organization, the Photographic Intelligence Division (PID), designed by Arthur Lundahl at CIA to create intelligence from the U-2 photography, changed the Cold War. Land would later envision the Corona film-return satellite, advise on the creation of the National Aeronautics and Space Administration (NASA) and the National Reconnaissance Office (NRO), and play a critical role in persuading President Richard Nixon to fund the Kennen (KH-11), the first digital electro-optical reconnaissance satellite.[6]

CHALLENGES TO DISCOVERY

As remarkable as the U-2's capabilities were, the Soviet Union was so vast that the U-2 could never photograph all of it. Accordingly, Allen Dulles, the Director of CIA created a committee to define what the U-2 should photograph. The Ad Hoc Requirements Committee on project AQUATONE (ARC) was chaired by James Q. Reber, a University of Chicago PhD and CIA officer, who had worked at the State Department during World War II. Reber's committee included representatives from the Army, Navy, Air Force, Department of State, CIA, and the National Security Agency (NSA). Previously, these organizations had agreed on which broad classes of targets would be collection priorities: Soviet offensive weapons systems and forces, weapon systems research and development programs, Soviet long-range aviation bases, major submarine bases, and key production bases for long-range bombers and submarines.[7]

In spite of agreement about the priorities, at ARC meetings on 1 and 4 June 1956, held to plan the initial U-2 missions, the entire intelligence community could locate only 45 intelligence targets on a map of the Soviet Union. As the meeting minutes record: "In the second meeting the 45 targets, having been plotted, discussion centered around the validity of some of the targets in terms of the relevancy to DCID 4/5."[8] Even before the first overflight of the Soviet Union, the ARC could not agree on the value of individual Soviet targets even when it knew only 45 of them.

Although intelligence about Soviet capabilities was critical, the White House focused equally on the political sensitivity of U-2 overflights of a nuclear-armed rival. While the ARC proposed targets, before each mission, the Director of Central Intelligence (DCI), Allen Dulles, would have to show proposed flight tracks and targets to President Eisenhower for his approval. As the President had intestinal surgery in May 1956, before the first Eastern Europe overflight missions, the initial U-2 mission conversation took place in Walter Reed Hospital. Finally, on 21 June, the President gave his approval and the first Soviet penetration overflight took place on 4 July 1956.[9]

The first Soviet U-2 penetration missions proved the U-2's value at seeing what had never been seen before. Lundahl's photointerpreters, who worked in a covert location over a car dealership in Washington, D.C., were the first Americans to look deeply into the Soviet Union. The first missions also revealed a critical operational gap in the U-2 program. While these flights showed how few strategic bombers the USSR had deployed, the program discovered that Soviet radars tracked each U-2 flight from end-to-end and that Soviet air defenses had made repeated attempts to shoot them down. American ignorance about Soviet radar capabilities increased both the pilot risk and political sensitivities. After 10 July, when the Soviet Union delivered a written protest to the US Embassy in Moscow with a map showing the U-2 flight tracks, the President halted Soviet penetration overflights. While these first U-2 overflights revealed great numbers of unknown military and industrial targets, and significant intelligence about Soviet bomber and submarine forces, they did not answer most questions about Soviet strategic missile and nuclear targets.[10]

In fall 1956 and spring 1957, no U-2 flew over the Soviet Union. The U-2s had been diverted in the fall of 1956 to cover the Suez crisis, while the nearly simultaneous Hungarian Revolution increased tensions between the West and the Soviet Union. The brutal Soviet armed repression in Hungary caused President Eisenhower to extend his order that no U-2 fly over the Soviet Union.[11] Throughout this interval, in the monthly ARC meetings, the Intelligence Community continued to request collection of potential Soviet nuclear and missile targets. In May 1957 President Eisenhower agreed to resume U-2 overflights, but even more than previously, he scrutinized each mission, and frequently changed flight tracks and deleted targets.[12] During this interval the U-2 had been modified to be less detectable by Soviet radars, but the modifications were ineffective. These early 1957 flights were restricted to the periphery of the Soviet Union. These peripheral flights did provide some important information about Klyuchi, on the Kamchatka peninsula north of Japan.

While much construction was ongoing there, so little was known about the Soviet missile test flights that the Intelligence Community could not determine if Klyuchi was a launch point or an impact area.[13]

The American uncertainty and ignorance about Soviet strategic missile developments was demonstrable. On 9 May 1957, Herbert Scoville, CIA's Assistant Director for Scientific Intelligence, wrote the Chairman of the ARC:

> Although we, like the other ARC members, have attempted to give the best possible guidance in our participation in the development of target lists, we feel there are certain subtleties relative to our targets which cannot be recorded in the confines of a standardized target list. I have in mind here such problems as the Novokazalinsk-Kyzl Orda-Dzusaly area, also Kyluchi, and our estimate of where missile or earth satellite activities may be going on there.[14]

What Scoville termed "subtleties," referred to the imprecise locations of all other intelligence sources, classified and unclassified, on this issue. The area Scoville referenced is slightly more than 220 miles long, or the distance between Washington D.C. and New York City, which illustrates the planning difficulties for the U-2 program. From other sources, US intelligence had been aware of activities in that area, but none of these sources could identify or locate precisely the facilities or activities.

DISCOVERY

The second intellectual activity of geospatial intelligence is discovery. The U-2 program had started overflying the Soviet Union in July 1956, but it had not yet discovered much intelligence about the Soviet Strategic Missile program. Consequently, in the summer of 1957, President Eisenhower approved a series of nine U-2 penetration overflights starting in August 1957. These nine missions were very successful and provided the first discovery of Soviet nuclear and missile facilities. In early August, US knowledge grew significantly when the missile test center at Tyuratam in Kazakhstan was first seen on U-2 photographs taken on 5 August 1957. The unprecedented size of the launch stand and the extent and amount of significant construction caused President Eisenhower to approve a second U-2 mission over Tyuratam on August 28.

The two photographs of the launch stand and its infrastructure were surprising and timely as Sputnik 1, the first successful space flight, was launched from Tyuratam on 4 October 1957. Each photograph provided the photointerpreters in Arthur Lundahl's Photo Intelligence Division new information about the Soviet missile program. The discoveries from the two U-2 missions over Tyuratam provided President Eisenhower information that helped him respond to the public clamor over the Sputnik launch. His response to the clamor was mostly silence as he had the intelligence provided by the U-2. But even after the discovery of the launch site, the photointerpreters and the Intelligence Community had many unanswered questions.

Discovery is the activity most subject to chance, as it carries an element of luck. Either or both August 1957 observations of Tyuratam could have been cloud-covered. The flight track of the first observation was south of the facility, and the second image was obtained from a different flight track. And whether important or routine, every geospatial discovery, is followed by the third intellectual activity—recording (Figure 1.1).

FIGURE 1.1   An enlargement of a U-2 photograph of the missile test stand and flame bucket at Tyuratam.

Source: CIA/PIC. Joint Photographic Intelligence Report, Missile Launching Complex and Test Range, Tyura Tam, U.S.S.R. HTA/ JR 4/58, September 1958. 6 (Approved for Release 2005/11/17, CIA-RDP02T6408R000900010026-3).

RECORDING

As soon as the photointerpreters at the Steuart building looked at the 5 August image, the recording and communicating of the new intelligence began. Recording is the third intellectual activity in geospatial intelligence. As unrecorded observations last only in the mind of the observer, this activity begins very quickly after discovery. The recording of the U-2 missions was sequential. The initial report described the mission accomplishments, and listed what targets were covered or photographed. This report began with a summary of what percent of the developed film was interpretable, and the body of the report was essentially a list of the sequential targets with the image number, and their latitudes and longitudes. For new targets, a brief physical description would be provided. For known targets, a count of equipment was noted. If a target was partially obscured by cloud or haze, or the photograph only covered part of a facility, that would be noted. As soon as the entire mission had been scanned, this report was completed and disseminated electronically through the US Intelligence Community as a text-only cable.

After the mission activity report was finished, a different recording process would commence. For existing targets, their target folder would be updated. The target folder was part of the information management process. Each target would be identified and all its information would be kept in individual manila folders. The text reports of the photointerpreters were kept there, along with any other associated information from other intelligence sources, such as maps, émigré reporting, or ground photography from Soviet sources or US defense attaches. The target folder formed the working files for the analysts and the compilation of information helped them determine how a target functioned or its level of activity.

Additionally, the photogrammetrists—mathematicians who specialized in measuring objects from photographs—were compiling another form of recording. They were calculating measurements of the new facilities observed on the two images, as well as measuring the distances between structures and the new objects, and the azimuths of the newly discovered antennas.

Each target had a unique number—the BE or Basic Encyclopedia number. The Basic Encyclopedia was a global data base, managed by the US Air Force. It was geographically organized and contained any target that might have military potential for attack in wartime. While the Basic Encyclopedia was stored in a computer, each organization that looked at aerial photography had its own local paper and film copies of the information.

For significant intelligence targets, like the launch test stand at Tyuratam, other forms of recording and communicating were created. Twenty-by-twenty-four-inch photographic enlargements of the test stand were created and mounted to present at briefings. Cartographers created local maps of the new installation and the ongoing construction to show the relation between the parts of the facility and the whole. All these records played a part in the analytic process, and many of them would be used in communicating the results of the analysis in briefings and reports. By August 1957, these recording practices were routine at HTAUTOMAT, the CIA cover name for the Photointerpretation Division, where Lundahl's interpreters worked over the car dealership. But the objects discovered at Tyuratam were not considered routine. The discovery and re-photographing of the ICBM missile test center at Tyuratam on 5 and 28 August 1957 would begin a systematic discovery of Soviet strategic missile capabilities, and would have long-lasting consequences for the US Intelligence Community.[15]

COMPREHENDING

After recording, comprehending is the next intellectual activity in geospatial intelligence. What distinguishes comprehending from discovery is the different questions that comprehending can answer. Discovery can answer often are questions about physical definition or what; questions of location or where; and questions of quantity or how many.

Comprehending can be defined as the movement from answering discovery questions to answering analytic questions such as how a facility or piece of equipment functions or why a facility or military force is configured in a certain way. But comprehending requires repeated relooking and comparing different images or photographs. It can also require putting the visual discoveries into an intellectual context.

On both U-2 photographs of Tyuratam even the best photointerpreters saw much that was new and incomprehensible. Art Lundahl, who directed HTAUTOMAT, briefed the initial results of these August missions to the Guided Missile Intelligence Committee (GMIC). Based on the intelligence discoveries by Lundahl's interpreters, and the great number of remaining questions about what had been seen on the images, the GMIC proposed that a "technical assessment by a group of selected, highly qualified US guided missile developmental and test personnel to be essential to fully exploit this AQUATONE data."[16] (AQUATONE was the CIA code name for the U-2 program.)

Lundahl agreed and on 24 September 1957 he wrote a memo that scheduled and organized the joint government–civilian analytic effort that he named the Jam Session.[17] Lundahl had designed an analytic experiment that would bring American civilian scientific and engineering experts into the Steuart Building, the covert location over the car dealership in a slum neighborhood.

JAM SESSION

The GMIC invited the foremost American guided missile experts to come and work with the PID photointerpreters in the Steuart building on analyzing the U-2 photography of the Tyuratam facility. Lundahl's briefing to the GMIC provoked the curiosity of the missile scientists and engineers as none of them had seen any facility resembling the Soviet launch test stand at Tyuratam. By 25 September, Lundahl had designed the working arrangements so that scientists and engineers could work side-by-side in concert with the support staff, the photogrammetrists, and other photointerpreters. A total of 26 ballistic missile and nuclear energy experts came,[18] among them Albert (Bud) Wheelon and Carl Duckett, both of whom would go on to lead CIA's Directorate of Science and Technology.[19] The 4 October launch of Sputnik 1, the first successful orbital satellite, from Tyuratam had increased their curiosity even more.

On the day after the second Soviet Sputnik launch from Tyuratam, 4 November 1957, the Jam Session began at the unlovely Steuart building. The two recent surprise Sputnik launches meant that the Jam Session analysis would occur covertly amid much publicity and debate about American national security (Figure 1.2).

Prior to the Jam Session, the photointerpreters' study of the two August U-2 images showed details that indicated the pace of the Soviet construction at Tyuratam:

> The observations were almost unbelievable. For example, at Communications Area B on the support base, on 5 August there were one double rhombic antenna array, one two-bay fishbone antenna, and one row of three stick masts. On 28 August, there were nine double rhombics, two fishbones, one three-mast array, one four-mast array, and three single masts. In those 23 days, 92 masts had been erected, an astonishing accomplishment.
>
> The feverish pace of construction indicated a crash effort to achieve operational readiness for the Center at the earliest possible date. Through the object of all this haste, whether for military advantage or some spectacular space event, was not apparent late in the summer of 1957, the reason became clear in the next several weeks. By the uninhibited use of a military booster, the Soviets were able to launch the first earth satellite ...[20]

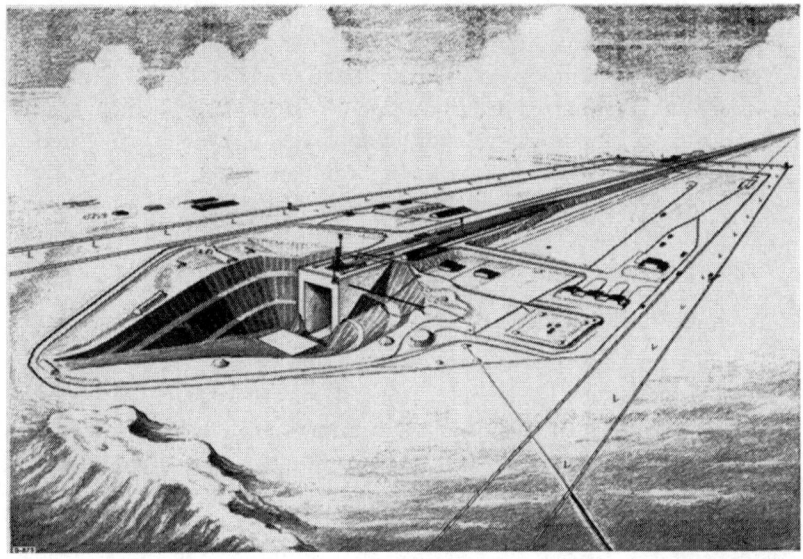

FIGURE 1.2 Artist rendering of the test stand and flame bucket at Tyuratam from the Jam Session report.

Source: CIA/PIC. Joint Photographic Intelligence Report, Missile Launching Complex and Test Range, Tyura Tam, U.S.S.R. HTA/ JR 4/58, September 1958. 6 (Approved for Release 2005/11/17, CIA-RDP02T6408R000900010026-3).

Before the Jam Session convened, the significance of the analysis and a rationale for the pace of activity had become evident. Sputnik 2, launched on 3 November 1957, differed significantly from Sputnik 1. Sputnik 2 was six times heavier than Sputnik 1. Its increased size and mass meant that the Soviet Union had solved the technical problems of launching into space on an intercontinental missile an object as heavy as a nuclear weapon.[21] The US had not yet solved these technical problems.

Amid the initial American political and public outcry about the two Sputniks causing the US to fall behind the USSR in the space race, the photointerpretation and technical analysis at the Jam Session began in secret. And within the Intelligence Community, the Jam Session provided three kinds of new comprehending:

Locational: Once the exact location of the test and launch site at Tyuratam had been learned, the discovery enabled researching of open-source material and tasking of other intelligence community sources to collect information about this site. The discovery also allowed analysis of Tyuratam in relation to other recently discovered sites, particularly Kyluchi. The detailed Jam Session analysis of the antenna construction and orientation at Tyuratam enabled the comprehension and confirmation that Kyluchi was the impact site for inter-continental ballistic missiles test-fired from Tyuratam.[22]

Analytical: The photography of the unique infrastructure discovered at Tyuratam allowed for measurement (or as photointerpreters call it, mensuration) of the launch tower, and a volumetric estimate of the excavated flame bucket. These measurements enabled the rocket scientists and engineers who came to the Steuart building to estimate the diameter, length, and thrust of the missiles that launched Sputnik 1 and Sputnik 2. Also, the rapid construction of antennas for missile telemetry provided a great deal of technical information for the missile experts as well as alerting the analysts about what activities might be forthcoming at Tyuratam.

Operational: The two U-2 photos of Tyuratam showed that the launch site components were connected by an unimproved road network, but the Soviet Union had done a great amount of construction to improve and extend the rail lines inside the facility. These observations indicated to the engineers and photointerpreters that the missile component transfers—movements of the payload and missile stages—at Tyuratam had to be done by rail. And from these observations, the Jam Session participants comprehended that Soviet intercontinental missiles were so large and heavy that a railroad was necessary for their transport.

TRACKING

Tracking is the fifth intellectual activity of geospatial intelligence. It is possible only after discovery and comprehending, and it requires observations at more frequent intervals. The frequency of observations required for tracking must exceed the pace of the activity being tracked. In the Jam Session analysis of Tyuratam, tracking was the only geospatial intellectual activity that was not possible. While the Jam Session provided the US with a key for future tracking of the Soviet Missile forces, there was insufficient photographic coverage for any tracking in 1957, and for many years to come.

This observation about the railroad tracks at Tyuratam turned out to be as important as the scientific and engineering analysis of the two unique objects, the large launch tower and flame bucket. At that time, the Soviet rail network covered only 60% of the Soviet land mass. Discovery of this essential dependency of Soviet missile transport on the rail network became a significant collection planning tool that bounded imagery collection for the U-2, and later the Corona satellites, for the next 15 years.[23] After the Jam Session, the US knew how to find and track the Soviet missile forces.

UNFORESEEN INTELLIGENCE COMMUNITY REACTIONS

While this new knowledge about Tyuratam was essential and valuable, it was not all well received inside the US Intelligence Community. The Jam Session results, published in preliminary form on 27 November 1957, disagreed with prior Intelligence Community assessments about Soviet missiles.[24]

The Jam Session forced CIA and other DoD all-source analysts to rapidly change their analysis and some of their assumptions. The combined efforts of the missile experts and the photointerpreters rendered invalid much previous Intelligence Community all-source analysis. For the first time, the work of the photointerpreters and outside industry experts, based on only two U-2 photographs, had provided undisputable information that forced CIA and DoD all-source analysts to rapidly revise their analysis and to recognize the amount of accurate information that could be gotten from analysis of aerial photography by photointerpreters and process experts.

While both organizations expected Lundahl's analysts to obtain new information, neither expected that it would produce intelligence sufficiently compelling that it would force the Intelligence Community to issue a new Special Intelligence Estimate on the Soviet ICBM program in December 1957.[25]

AN UNINTENDED BENEFIT

The Jam Session resulted in an additional access to communications intelligence for the photointerpreters. The observations and azimuth mensuration of all the antennas at Tyuratam, based on the type, orientation, and locations of the antennas that collected telemetry, enabled the analysts to establish a relation between Kyluchi and Tyuratam. As a consequence of the Jam Session analysis, the PID Military-Scientific Branch was granted access to communications intelligence (COMINT) from NSA.[26] The combination of locational intelligence obtainable only from overhead imagery with human intelligence and communications intelligence

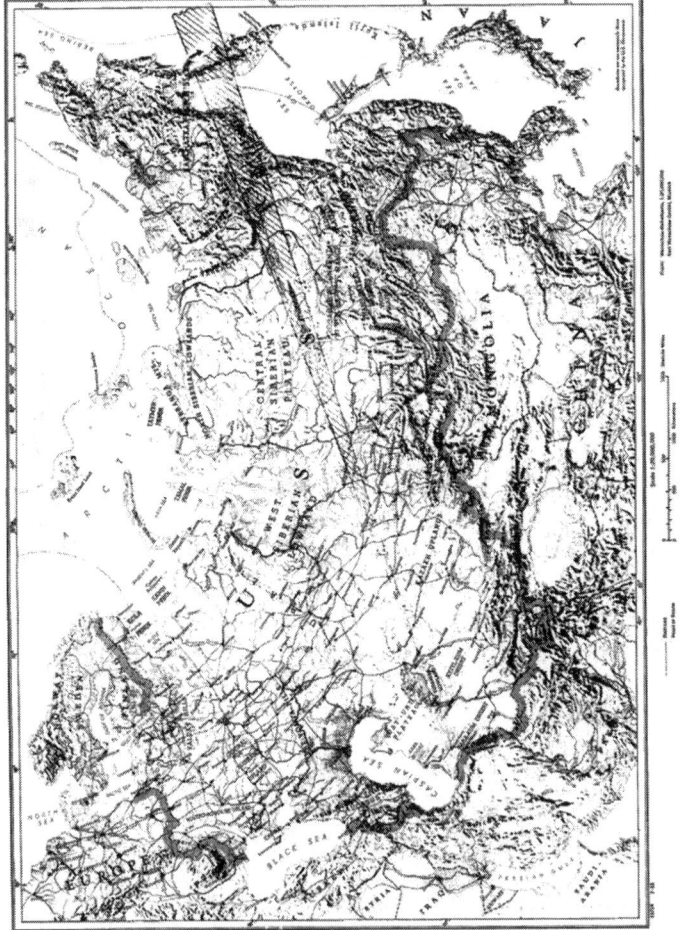

FIGURE 1.3   Map of the Soviet Union from the Jam Session report showing relation of Tyuratam to Kyulichi.

Source: CIA/PIC. Joint Photographic Intelligence Report, Missile Launching Complex and Test Range, Tyura Tam, U.S.S.R. HTA/JR 4/58, September 1958. 6 (Approved for Release 2005/11/17, CIA-RDP02T6408R000900010026-3).

facilitated the beginning of the mapping of Soviet strategic missile infrastructure (Figure 1.3).

The US, due to Din Land, was aware of the U-2 vulnerability to shootdown even before its creation, and since 1957 had been busily designing and building a photographic imaging satellite.[27] By December 1957, the risk assessment in the U-2 program was sufficiently high that only six more Soviet penetration overflights were attempted: one in 1958, two in 1959, and three in 1960. On the last one, the Soviets shot down a U-2 and captured Francis Gary Powers, its pilot. Through the remainder of the U-2 program, Tyuratam was photographed only three more times, in July and December 1959 and April 1960.[28] Three months after the last, failed U-2 mission, the first Corona imaging satellite launch and recovery occurred on 18–20 August 1960.[29]

The combination of the foremost US missile experts—with backgrounds in physics and engineering, and with access to communications intelligence and human intelligence—and Lundahl's photointerpreters—empiricists *par excellence*—resulted in intelligence discoveries that neither group could have achieved on its own. The Jam Session's combination of photointerpretation, antenna analysis allowing the relation of Tyuratam to Kyluchi, and the initial mapping of the Soviet Strategic infrastructure was unprecedented. The map in the Jam Session report indicates the state of US government mapping in 1958 as the Jam Session intelligence was overlaid on a German commercial map of the USSR.

For the first time in US intelligence history, all the disciplines later united under the 2005 legal definition of GEOINT in U.S.C. § 467 US Code 467—"the exploitation and analysis of imagery and geospatial information to describe, assess, and visually depict physical features and geographically referenced activities on the earth" had been combined in one analytic effort. The Jam Session analytic experiment also is the archetypal illustration of four of the five intellectual activities—envisioning, discovery, recording, and comprehending—involved in the practice of geospatial intelligence. It also provided a key that would enable the fifth intellectual activity—tracking—in the late 1970s.

In the 48 years between the Jam Session and the legal definition of GEOINT, these different analytic activities took multiple organizational forms in CIA, DIA, military service intelligence centers, US Defense Mapping Agency (DMA), and the National Photographic Interpretation Center (NPIC), but no prior event in American intelligence history comes close to the definition of geospatial intelligence in conception, execution, or consequences. It is arguably the genome of GEOINT (geospatial intelligence), and the following chapters of the book will examine the history of the intellectual activities that comprise geospatial intelligence and their future implications.[30]

NOTES

1 Evan Thomas, *Ike's Bluff: President Eisenhower's Secret Battle to Save the World* (NY: Little Brown, 2012), p. 181; Michael Bechloss, *Mayday: Eisenhower, Khrushchev and the U-2 Affair* (NY: Harper and Row, 1986), p. 119.

2 Killian, James R. *Sputnik, Scientists, and Eisenhower: A Memoir of the First Special Assistant to the President for Science and Technology* (Cambridge, MA: MIT Press, 1977), pp. 79–85. The panel was informally called the Killian Panel.

3 Beacon Hill Report. Problems of Air Force Intelligence and Reconnaissance, Project Lincoln, Massachusetts Institute of Technology (MIT), 1952. www.governmentattic.org/12docs/USAF-BeaconHillReport_1952.pdf.

4 Taubman, Phillip. *Secret Empire: Eisenhower, the CIA, and the Hidden Story of America's Space Espionage* (New York, NY: Simon & Schuster, 2003), pp. 78–85.

5 Land, Edwin H. Memorandum for Director of Central Intelligence, "A Unique Opportunity for Comprehensive Intelligence," 5 November 1954, Document 3, https://nsarchive2.gwu.edu/NSAEBB/NSAEBB54/. Viewed 4 September 2022.

6 Bahcall, Safi. *Loonshots: How to Nurture the Crazy Ideas that Win Wars, Cure Diseases, and Transform Industries* (NY: St. Martin's Press, 2019), pp. 113–116; Center for the Study of National Reconnaissance. *The Era before KENNAN/KH-11.* Volume I, 55–64. (Approved for Release 2021/07/09 C05097836), www.nro.gov/Portals/65/docume nts/foia/declass/HISTORICALLY%20SIGNIFICANT%20DOCs/NRO%2060th%20Anniversary%20Docs/SC-2021-00002_C05097 836.pdf. Viewed 18 September 2022.

7 Pedlow, Gregory W. and Donald E. Welzenbach, *The CIA and the U-2 Program*, pp. 80–81; Intelligence Vital to National Security through AQUATONE TB# 143488/B (Approved for Release 2006/02/07 CIA-RDP92B01090R002500010080-2). Project AQUATONE was the Code name for the U-2. Pedlow and Welzenbach, p. 40.

8 CIA. AD HOC REQUIREMENTS COMMITTEE ON PROJECT AQUATONE (ARC) Minutes of Meetings Held in Room 121 Administration Building, Central Intelligence Agency, at 1200, 1 June 1956 and at 1400, 4 June 1956. SAFC 678 (Approved for Release 2000/08/21/ RIA-RDP33-02415A000100100056-7). DCID 4/5 defined the intelligence priorities for the U.S.

9 Pedlow and Welzenbach, p. 100.

10 CIA National Photographic Interpretation Center. *The Years of Project HTAUTOMAT 1956–1958,* Vol I and II, December 1974 (Sanitized Copy Approved for Release 2010/03/12, CIARTP04T00184R000400010001-1), pp. 107–08.

11 Pedlow and Welzenbach, p. 123.

12 Pedlow and Welzenbach, p. 128.

13 CIA. ARC and Reber memos about Kyzl Orda. 13 May 1957, Memo from Chairman Ad Hoc Requirements Committee to Director of Operations, subject, Ad Hoc Requirements Committee requirements for the KLYUCHI Mission. (Approved for Release 2003/12/23: CIA_RDP61S00750A00050003117-3), TCS-1411-57, ARC-M-43. Special Meeting of the Ad Hoc Requirements Committee on Project AQUATONE (ARC), 19 May 1957; 22 May 1957, SC-03325-57 Memorandum for Ad Hoc Requirements Committee Members Subject Detailed Requirements for Penetration Flights (Approved for Release 2001/08/09 CIA-RDP61S00750A000200070022-7;TCS-1506-57, ARC-M-46), Ad Hoc Requirements Committee on Project AQUATONE (ARC), 27 May 1957. (Declassified in Part-Sanitized Copy Approved for Release 2013/08/28: CIA-RDP61S00750A000200070013-7); Draft: J.Q. Reber, 10 July 1957 (Declassified and Approved for Release 2013/09/10: CIA-RDP61S00750A000200080106-3); Targets for Coverage in General Area of 4035 (no date). Approved for Release 2003/12/23CIARDP61S00750A000500030098-5.

14 Scoville, Herbert, Jr. "Memorandum for Chairman, Ad Hoc Requirements Committee, Subject: Flight Planning for Future Aquatone Missions," 9 May 1957. (Sanitized Copy Approved for Release 2003/12/23: CIA-RDP61S00750A000500030122-7).

15 CIA. National Photographic Interpretation Center. *The Years of Project HTAUTOMAT 1956–1958,* Vol I and II, December 1974 (Sanitized Copy Approved for Release 2010/03/12, CIARTP04T00184R000400010001-1), pp. 159–162.

16 Quote taken from citation in 3 October 1957 Memo for Director of Central Intelligence, Subject: *Production of Guided Missile Intelligence from Evidence Gathered by AQUATONE* from Herbert Scoville, Jr. Assistant Director Scientific Intelligence. (Approved for Release Date 15 Jul-2011 (Scoville Memo)). Pedlow and Welzenbach, p. 40.

17 CIA. National Photographic Interpretation Center. *The Years of Project HTAUTOMAT 1956–1958,* Vol I and II, December 1974 (Sanitized Copy Approved for Release 2010/03/12, CIARTP04T00184R000400010001-1), pp. 186–188.

18 The Jam Session also analyzed a number of Soviet nuclear energy targets that had been collected on the August U-2 missions, but that analysis remains classified.

19 Dietel, Ed. "Charting a Technical Revolution: An Interview with Former DDS&T Albert Wheelon," *Studies in Intelligence,* 1989 (Approved for Release: 2014/09/10 C00863247) pp. 32–34.

20 CIA. National Photographic Interpretation Center. *The Years of Project HTAUTOMAT 1956–1958,* Vol II, December

1974 (Sanitized Copy Approved for Release 2010/03/12, CIARTP04T00184R000400010001-1), p. 213.

21 "Sputnik 1 and Sputnik 2," Wikipedia. https://en.wikipedia.org/wiki/Sputnik_1 and https://en.wikipedia.org/wiki/Sputnik_2, viewed 30 October 2017.

22 Initally, the assessment was that the Tyuratam missile test center covered about 40 square miles. (CIA. Guided Missile Intelligence Committee, Report of the Special Engineering Analysis Group, 27 November 1957, Washington D. C. (Approved for Release 2012/05/01: Cia-RDP78T05439A000300350035-6), p. 8. By 1964, the US knew that the missile test center at Tyuratam covered 1,500 square miles, an area approximately the size of Rhode Island. CIA. Scientific Intelligence Report. *New Space Facilities at the Tyuratam Missile Test Center.* 14 October 1964. Preface, Page v. (Approved for Release 2003/09/02: CIA-RDP78T05439A000400060010-4).

23 CIA. National Photographic Interpretation Center. *The Years of Project HTAUTOMAT 1956–1958,* Vol II, December 1974 (Sanitized Copy Approved for Release 2010/03/12, CIARTP04T00184R000400010001-1), p. 248.

24 CIA. Guided Missile Intelligence Committee. *Report of the Special Engineering Analysis Group.* 27 November 1957, Washington D.C. (Declassified in Part—Sanitized Copy Approved for Release 2012/05/01: CIA-RDP78T05439A000300350035-6).

25 CIA. Special National Intelligence Estimate Number 11-10-57. The Soviet ICBM Program. 17 December 1957 (Approved for Release 15-Jul-2011).

26 CIA. National Photographic Interpretation Center. *The Years of Project HTAUTOMAT 1956–1958,* Vol III, December 1974 (Sanitized Copy Approved for Release 2010/03/12, CIARTP04T00184R000400030001-9), pp. 356–357.

27 But the technical challenges of building and flying that satellite, Corona, emerged with its first failed launch on 21 January 1959, 25 months after the Jam Session ended. Greer, Kenneth E. "Corona: The First Photographic Reconnaissance Satellite," *Studies in Intelligence,* Supplement 17 (Spring 1973): 1–37, in *Corona: America's First Satellite Program* (Washington, D.C.: CIA Center for the Study of Intelligence History Staff, 1995), pp. 6 and 12.

28 Pedlow and Welzenbach, p. 143; *New Launch Area and Other Developments, Missile Launching Complex Tyuratam, USSR: A Comparative Analysis of 1957, 1959, and 1960 Photographs.* PIC/JR-17/60, July 1960 (Declassification Review by NIMA/DoD, HR70-14, Approved for Release 16 Jul 2010), p. 26.

29 Ruffner, Kevin. C. ed. *Corona: America's First Satellite Program* (Washington, D.C.: CIA Center for the Study of Intelligence History Staff, 1995), p. 1.

30 10 U.S.C. § 467 *US Code—Section 467: Definitions; Memorandum for Principal Director of National Intelligence, Deputy Director of National Intelligence for Collection,* from James R Clapper, Lieutenant General USAF (Ret), Director (NGA) 15 October 2005. The de jure definition of Geospatial Intelligence is "geospatial intelligence" means the exploitation and analysis of imagery and geospatial information to describe, assess, and visually depict physical features and geographically referenced activities on the earth. Geospatial intelligence consists of imagery, imagery intelligence, and geospatial information.

BIBLIOGRAPHY

Bahcall, Safi. *Loonshots: How to Nurture the Crazy Ideas that Win Wars, Cure Diseases, and Transform Industries* (NY: St. Martin's Press, 2019).

Beacon Hill Report: Problems of Air Force Intelligence and Reconnaissance, Project Lincoln, Massachusetts Institute of Technology (MIT), 1952.

CIA. AD HOC REQUIREMENTS COMMITTEE ON PROJECT AQUATONE (ARC) Minutes of Meetings Held in Room 121 Administration Building, Central Intelligence Agency, at 1200, 1 June 1956 and at 1400, 4 June 1956. SAFC 678 Approved for Release 2000/08/21/ RIA-RDP33-02415A000100100056-7.

—— Intelligence Vital to National Security Through AQUATONE TB# 143488/B (Approved for Release 2006/02/07 CIA-RDP92B01090R002500010080-2).

——National Photographic Interpretation Center. *The Years of Project HTAUTOMAT 1956–1958,* Vol I and II, December 1974 (Sanitized Copy approved for release 2010/03/12, CIARTP04T00184R000400010001-1).

—— National Photographic Interpretation Center. *The Years of Project HTAUTOMAT 1956–1958,* Vol III, December 1974 (Sanitized Copy approved for release 2010/03/12, CIARTP04T00184R000400030001-9).

——13 May 1957, Memo from Chairman Ad Hoc Requirements Committee to Director of Operations, subject, Ad Hoc Requirements Committee requirements for the KLYUCHI Mission. (Approved for Release 2003/12/23: CIA_RDP61S00750A00050003117-3), TCS-1411-57.

——ARC-M-43 Special Meeting of the Ad Hoc Requirements Committee on Project AQUATONE (ARC), 19 May 1957.

——ARC-M-46, Ad Hoc Requirements Committee on Project AQUATONE (ARC), 27 May 1957. (Declassified in Part-Sanitized Copy Approved for release 2013/08/28 CIA-RDP61S00750A000 200070013-7).

————Draft: J.Q. Reber, 10 July 1957 (Declassified and Approved for Release 2013/09/10: CIA-RDP61S00750A000200080106-3; Targets for Coverage in General Area of 4035 (no date) Approved for release 2003/12/23CIARDP61S00750A000500030098-5.

————Guided Missile Intelligence Committee, Report of the Special Engineering Analysis Group, 27 November 1957, Washington D. C. (Approved for Release 2012/05/01: Cia-RDP78T05439A000300350035-6).

————Scientific Intelligence Report. *New Space Facilities at the Tyuratam Missile Test Center.* 14 October 1964. Preface, Page v. (Approved for Release 2003/09/02: CIA-RDP78T05439A000400060010-4).

———— Special National Intelligence Estimate Number 11-10-57. The Soviet ICBM Program. 17 December 1957. (Approved for Release 15-Jul-2011).

CIA/PIC. Joint Photographic Intelligence Report, Missile Launching Complex and Test Range, Tyura Tam, U.S.S.R. HTA/JR 4/58, September 1958. 6 (Approved for Release 2005/11/17, CIA-RDP02T6408R000900010026-3).

Dietel, Ed. "Charting a Technical Revolution: An Interview with Former DDS&T Albert Wheelon," *Studies in Intelligence,* 1989 (Approved for Release: 2014/09/10 C00863247).

Killian, James R. *Sputnik, Scientists, and Eisenhower: A Memoir of the First Special Assistant to the President for Science and Technology* (Cambridge, MA: MIT Press, 1977).

Land, Edwin H. Memorandum for Director of Central Intelligence, "A Unique Opportunity for Comprehensive Intelligence," 5 November 1954, Document 3, https://nsarchive2.gwu.edu/NSAEBB/NSAEBB54/.

National Reconnaissance Office (NRO) Center for the Study of National Reconnaissance. *The Era before KENNEN/KH-11.* Volume I, 55–64 (Approved for Release 2021/07/09 C05097836). www.nro.gov/Portals/65/documents/foia/declass/HISTORICALLY%20SIGNIFIC ANT%20DOCs/NRO%2060th%20Anniversary%20Docs/SC-2021-00002_C05097836.pdf.

Pedlow, Gregory W. and Donald E. Welzenbach, *The CIA and the U-2 Program.* (Washington, D.C.: Center for the Study of Intelligence History Staff, 1998).

Ruffner, Kevin. C. ed. *Corona: America's First Satellite Program.* (Washington, D.C.: CIA Center for the Study of Intelligence History Staff, 1995).

Scoville, Herbert, Jr. "Memorandum for Chairman, Ad Hoc Requirements Committee, Subject: Flight Planning for Future Aquatone Missions," 9 May 1957 (Sanitized Copy approved for Release 2003/12/23: CIA-RDP61S00750A000500030122-7).

"Sputnik 1 and Sputnik 2," Wikipedia. https://en.wikipedia.org/wiki/Sputnik_1 and https://en.wikipedia.org/wiki/Sputnik_2.

10 U.S.C. § 467 *US Code—Section 467: Definitions; Memorandum for Principal Director of National Intelligence, Deputy Director of National Intelligence for Collection*, from James R Clapper, Lieutenant General USAF (Ret), Director (NGA), 15 October 2005.

CHAPTER 2

Envisioning

INTRODUCTION

The first characteristic of geospatial intelligence is envisioning, a form of seeing differently.[1] Because envisioning combines the visual with the cerebral, it demonstrates the abilities of some humans to see differently from others. This internal form of envisioning combines, in the service of problem solving, the sensory trait of visual acuity with some combination of eidetic memory and insight. This quality is captured in the idiom "the mind's eye." It often results in a new way of seeing the world or part of the world that diminishes the effect of incomplete or ambiguous information.

When man incorporates technology to see differently, a more modern form of envisioning results. The examples of mechanical envisioning, often created by discoverers, found their way into solving geographic problems. The development of quadrants, sextants, telescopes, and microscopes illustrates this principle, but the use of the stereographic aerial photography in resolving terrain features in World War I provides the best example of mechanical envisioning.

One 20th-century man had three visions related to photography: the first made his fortune; the second helped end the Cold War, and the third transformed photography. In 1943, in response to a question from his young daughter, Edwin (Din) Land envisioned instant chemical photography in sufficient detail to patent the technology that would become the Polaroid Land Camera. In 1952, Land envisioned the aircraft and camera that would become the U-2 reconnaissance plane four years later. And in 1969, Land envisioned the potential for digital imagery and persuaded the US government to invest in near-real-time imaging satellites for space reconnaissance, the first of which went into space in 1976.

Digital envisioning enables nearly all humans to use digital images, whether terrestrial, aerial, or space-based. It is an outgrowth of precision digital imagery and geographic information systems (GIS). The creation

 DOI: 10.4324/9781003436836-3

of Google Earth enabled digital envisioning and rudimentary geospatial analysis for all and brought sophisticated geospatial analysis to many. Google provided the envisioning tools that make the other characteristics of geospatial analysis—discovery, comprehending, tracking, and recording—possible for most people with connectivity.

IMOLA, OR HOW DA VINCI TAUGHT A TYRANT TO SEE DIFFERENTLY

The earliest example of geospatial thinking dates back to 1502, and it helped solve a military engineering problem in northeastern Italy, in the region between Bologna and Ancona called the Romagna. Cesare Borgia, bastard son of Cardinal Rodrigo Borgia, began the last of his three campaigns to seize this region and to expand the territory under his control. Borgia earned his reputation as a cunning offensive military leader. He was also aware of the difficulty of defending terrain. In particular, Borgia was wary of the French horse-drawn cannon that had broken open several walled Italian towns once thought to be safe from sieges. Borgia wanted a military engineer to help him defend the town of Imola.

Frequently, artists teach other people how to see differently. Once others have been taught, the artist's initial vision can turn into a new way of seeing. Cesare Borgia, on Niccolò Machiavelli's recommendation, hired Leonardo Da Vinci as his military engineer.[2] Under this charge, Da Vinci was the first to create an aerial map from terrestrial measurements. He drew the Italian town of Imola from a bird's eye perspective around 1502. Da Vinci created this map to make Borgia see the town differently and to consider the elevation of the surrounding region. The map accurately captures the changes in the streets and the spaces between and among the structures.[3]

Leonardo's map is the first example in the west that combined an artistic envisioning capability with the mathematical understanding of spatial relations necessary to create an accurately measured aerial map or bird's-eye view. His 1502 map of Imola was the first that captured what the Roman Vitruvius could envision, an aerial view of a city plan. Unlike many of his other drawings and paintings, Leonardo did not create this map for scientific or aesthetic reasons. As a military engineer, Da Vinci created the map to improve the defensive earthworks around the town. To make his plan effective, before he created the map of Imola, DaVinci devised a tool to measure accurately the length and width of the streets and structures of Imola. Leonardo invented the hodometer,[4] a measuring device based on counting the revolutions of a wheel of known circumference. From his compilation of measurements, Leonardo was able to calculate the dimensions of the streets and structures and connect them to control points to keep his map as accurate as possible.

Leonardo's hodometer marks a starting point toward achieving an important attribute of geospatial intelligence—precision location. By creating a tool to measure accurately the distances within Imola, Da Vinci combined his envisioning intelligence with a new technology. The result provided Cesare Borgia an information advantage against a future siege. A comparison of Da Vinci's map to a modern Google Earth satellite image of Imola shows that the artist was able to capture more mathematical accuracy with a primitive tool to measure distances and angles, and careful measurement by pace, than many who followed him.

Leonardo demonstrated his ability to "see" from the air and other drawings captured his desire to travel in the air. His more commonly known drawings of many of his "flying machines" were also among the first to envision how humans would be able to routinely observe the earth from this point of view. In his plan to defend Imola, Leonardo was among the first to combine art and mathematics to create a spatially accurate map. But Leonardo could only extend precision measurements to the walls of Imola. His estimates of distances to the nearest towns, also recorded on his map, within his drawn circle but outside the town, were less precise than his measurements inside the town. While Leonardo could use mathematics and his hodometer to improve the precision of his pacing in Imola, he could not extend the precision of his measurements to the countryside.

It would take a long time before the precision measurements Da Vinci achieved at Imola could be extended outside the town walls. The quest for precision would require a number of new technologies—the telescope, the theodolite, the quadrant, sextant, and the chromometer—before other European scientists, inventors, and surveyors could achieve or extend Da Vinci's level of accuracy to other locations. And before that could happen, the mathematical work necessary to solve location problems was the first challenge for his successors.[5] As in Leonardo's case, their motivation to solve these problems was not abstract, but driven by the awareness of potential conflict,[6] trade, and the desire of European nations to discover and explore more of the world to increase their wealth. Da Vinci combined his artistic envisioning capability with his interest in mathematics[7] to envision a new way of looking at a military problem (Figure 2.1).

From the 16th century through the late 18th century, mathematical discoveries enabled increasingly accurate measurements of larger areas of the earth's surface. These mathematical discoveries inspired the instrument makers of that time to improve their quadrants, sextants, telescopes, compasses, and time pieces. Slightly more than 100 years after Da Vinci,

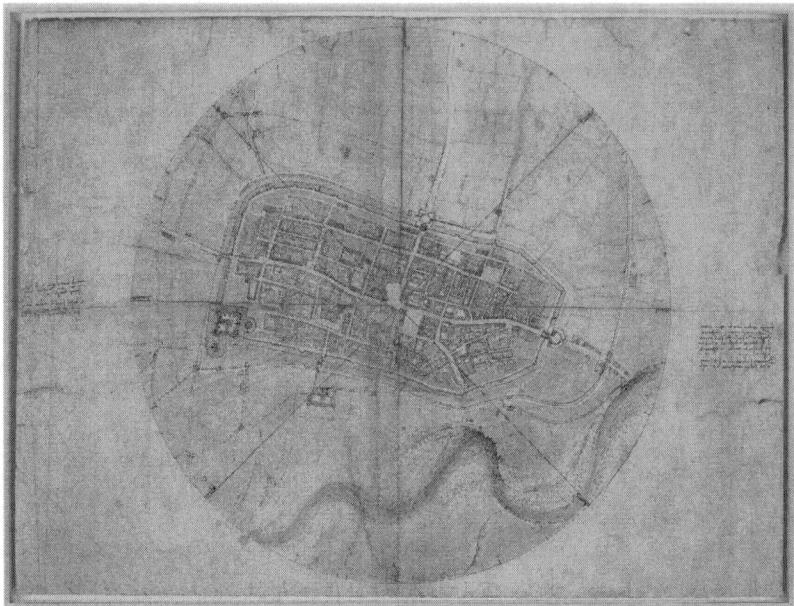

FIGURE 2.1 Leonardo da Vinci's 1503 bird's-eye-view map of Imola.
Royal Collection Trust/© His Majesty King Charles III 2023.

Galileo's work on telescopes[8] in the first years of the 17th century along
with the improvement of the sextants helped refine the centuries-old process of celestial navigation that measured latitude. It would take another
160 years for John Harrison to develop a clock that would keep time
accurately enough and long enough to enable accurate readings of longitude at sea to complement the accurate readings of latitude.[9] And
throughout this time, the human motivation to be able to rise above the
earth to see farther and better continued to develop.

In this era characterized by national explorations and the rise of
European states, nations began to commission mapmakers to help define
borders, understand the vulnerabilities of neighboring states, and pro-
mote research and exploration for conquest, religion, and trade. It would
take until the 17th century for surveyors to begin to use the mathematics of triangulation along with theodolites and survey chains to extend
accurate mapping through much of Europe. The work of Picard and the
Cassinis in France and William Roy and Jesse Ramsden in England drove
the creation of national cartographic organizations, and their output was
driven by their respective national military needs.[10] Much of the mapping
was done by military engineers, Leonardo's descendants, who sought how

to use or compensate for elevation, terrain, and the relative positions of natural obstacles.

ELEVATION AND TECHNOLOGY, OR SEEING LIKE DA VINCI

While developments in human ability to measure elevation had continued since Leonardo's time, only in 1783 in France with the invention of the hot-air balloon could someone else see the earth as Leonardo envisioned at Imola, 281 years before. After the first balloon ascent in France, it did not take long for others to envision a military purpose for this new technology. Eleven years later, during the battle of Fleurus, the French army used a tethered balloon for observation in the first military application of this technology, but the French observations were not recorded. This initiative began the development of the means for military aerial observation in combat and later, capturing these observations visually.[11]

ELEVATION AND PHOTOGRAPHY—SEEING DIFFERENTLY FROM DA VINCI

Two words summarize the earliest days of aerial photography: risk and failure. James Black took the oldest surviving aerial photograph 76 years after the battle of Fleurus, in 1860, from a balloon at 2,000 feet over Boston, Massachusetts. It was his only success among eight attempts on that particular day. Part of the challenge for Black was the fact that he was using heavy glass plates. The second challenge, according to contemporaneous reports, was the wind which affected the stability of the balloon and his photographic equipment.

A few years later, during the American Civil War, Thaddeus Lowe, an inventor, self-taught chemist, and self-promoter attempted to influence the Union Army to use balloons for aerial observation, both for artillery spotting and reconnaissance. Undocumented attempts were made to use balloons to take photographs from the air. Lowe had some success in reconnaissance, but he was unsuccessful as a photographer and he had a negligible military influence on the Union Army.[12] Yet, Lowe's most consequential act might have been sharing information about balloons with a Prussian military observer attached to McClellan's campaign. Count Von Zeppelin had his first balloon ride in Minnesota after that campaign[13] before returning to Prussia. After Lowe's experiments, the US military did not use or think seriously about balloons until many years later. Later in the 19th century, in Germany and the US, experimental attempts to attach aerial cameras to kites, homing pigeons, and even primitive rockets all produced results, but none that could be successfully replicated. Throughout the end of the 19th century, the efforts to replicate Leonardo's achievement at Imola still were unsuccessful. Yet, the terrestrial work of measuring and mapping areas of the earth had progressed significantly.

Unlike balloons, two networked civilian technologies—telegraphy and railroads—did play significant parts in the American Civil War and changed the nature of subsequent European wars. Both of these networks depended on accurate surveying and mapping for their effectiveness. They also introduced the need for more precise timekeeping in military operations. After the US Civil war, the commercial growth in the US and Europe impelled governments and businesses to map uncharted regions initially and charted regions more accurately. The military forces of European and North American nations played a significant part in increasing the quantity and accuracy of the mapping. However, throughout the 19th century, their technology was all ground-based and the mapping work was all accomplished with surveying instruments. While aerial photography remained experimental, terrestrial photography continued to develop steadily, and the early development of stereo-photography and motion photography took place.

In World War I when aircraft became weapons of intelligence as well as destruction, most of the combatant nations would use aerial photography on an industrial scale. After the initial Italian use of aircraft photography in 1911 during the Italo-Turkish war, military intelligence and cartographic efforts in World War I complemented the duplication, industrialization, and incremental improvements in aerial photography with photointerpretation and analysis.[14] As in the 17th and 18th centuries, technologies that enabled increased precision and understanding continued to be invented and improved. Among some World War I military leaders, the envisioning capability to make use of these tools also progressed. But World War I introduced a human limitation for some that continues to affect geospatial intelligence. This technology does not always produce clear results for all.

Fifty years after the first successful use of aerial photography for military intelligence, the 1962 Cuban Missile Crisis illustrates a perpetual challenge faced by those who can envision. While some people can see from a bird's-eye point of view, others cannot. Some people cannot translate their individual point of view into a different perspective. When those who can see from above attempt to inform those who cannot, the attempt can be challenging. The most significant example of this occurs in Bobby Kennedy's memoir of the Cuban missile crisis.[15] On his initial look at what the photointerpreters at CIA's National Photographic Interpretation Center identified on the ground in Cuba from U-2 aerial photography, Kennedy said:

Experts arrived with their charts and pointers and told us that if we looked carefully we could see there was a missile base being constructed in a field near San Cristobal, Cuba. I, for one, had to take their word for it. I examined the pictures carefully, and what I saw appeared to be no more than the clearing of a field for a farmer or the basement of a house. I was relieved to hear later that this was the reaction of virtually every one at the meeting, including President Kennedy. Even a few days later, when more work had taken place at the site, he remarked that it looked like a football field.

An involved, but not present, participant in the photo-intelligence support, Dino Brugioni, questioned Robert Kennedy's account.[16] Yet Robert Kennedy's admission about not seeing or recognizing details on a magnification of an image taken from 14 miles above the earth does not seem to be written to embellish his or anyone else's part in the Cuban Missile Crisis.

Robert Kennedy's comment about not being able to see objects on aerial imagery applies to some portion of the global population. Some can envision a scene and detect, seemingly intuitively, when an object is missing or mis-located, no matter the orientation or point of view. Other humans, over the course of time, develop unique abilities to perceive distinctions that many others have missed. The envisioning ability has enabled individuals to make significant human contributions in many areas of human achievement, including medicine, crime solving, geographic and geologic research and, like Leonardo Da Vinci, military intelligence. Yet, Robert Kennedy's experience illustrates the perceptual challenge for those who can envision. Whether it is diminished visual acuity, a lack of visual imagination, or a deficiency in spatial thinking, practitioners of geospatial intelligence would do well to remember that sometimes their clarity can be someone else's confusion or incomprehension. Some people lack any envisioning capability. The near universal access to images from aircraft and satellites in the 21st century may diminish the scale of the communicative challenge for those who can envision, but this communications challenge will not disappear entirely.

UBIQUITOUS ENVISIONING, OR ENABLING NEARLY EVERYONE TO SEE AS DA VINCI DID

National security caused the opposite envisioning experiences of Dino Brugioni and Bobby Kennedy during the Cuban Missile Crisis in 1962 to be equally rare. At that time, only select individuals in a few national government intelligence communities or military components were permitted to see photographic intelligence from aircraft and satellites. Fewer individuals were allowed to make it. But 40 years later, with the creation of Google Earth, both Kennedy's and Brugioni's experiences were made globally universal.

The transformational changes brought about by the spread of the internet in the late 1990s and early 2000s diminished and enlarged our perceptions of geography. The constraints of geography had already been diminished by the speed at which humans could send textual information across the country and around the planet. Internet traffic began with verbal communications, but the need to transmit vocal, visual, and video information led to the proliferation of multiple file formats and the need for tools to search the newly accessible troves of information. The need to search for or invent useful visual search tools led to competition among the Internet search engines—Google, Netscape, Microsoft Explorer, and others, as well as the digital GIS companies, ESRI, Intergraph, and others.

In the competition for market share, Google beat other search engines. One part of Google's success evolved from its unconstrained approach to autonomous employee research and development. Out of the Google culture, which allowed a half-day every week for employee experimentation, a number of Google employees made an internal case for developing a spatial searching capability. Their internal analysis of Google search inquiries revealed that 25% of Google queries began with the same word, "where," and at that time, Google did not have a map as part of its display of search information.[17]

From this internal analysis, Google investigated other companies doing spatial search and found a company called Keyhole. This company, on account of its ability to combine and geospatially register different satellite and aircraft images from different sources, already had other large organizations interested in its work. A newly created[18] commercial investment arm of CIA, In-Q-Tel, on behalf of the National Imagery and Mapping Agency, invested in Keyhole in spring 2003 to sustain and encourage its development. As Keyhole grew, it hired John Rohlf, who transformed its capability with the creation of the Keyhole Markup Language (KML). This language allowed every user of this software to share geospatial data, and additional data such as measurements and annotations. The global media incorporated Keyhole technology in its coverage of the 2003 Iraq war. Google, among a number of other interested companies that were trying to purchase Keyhole, bought the company outright in October 2004.[19] Google used and developed Keyhole technology to launch Google Earth and Google Maps, free services that allow anyone with internet access to look at aerial or satellite imagery and discover if his or her experience is more like Robert Kennedy's or Dino Brugioni's.

FIGURE 2.2   Google Earth satellite image of Imola.

Map data: Google.

Google Earth took advantage of the growth of the commercial satellite imagery that occurred in the late 1990s and early 2000s. Google's
introduction of Street View in May 2007,[20] combined ground view
images at geospatially registered locations with the aerial or satellite view
of the identical locations. Street View eased the challenge of envisioning
for those who were more like Bobby Kennedy. Google Earth technology
enabled anyone in the 21st century to see the same views of Imola as
Leonardo DaVinci envisioned from the air and Cesare Borgia saw from
the streets in 1502. Remarkably, Da Vinci's, Borgia's, and our views of
Imola are now similar. The same streets can be walked; the same roads are
visible, and the accuracy of Leonardo's envisioning can be observed and
measured today by anyone on a home computer (Figure 2.2).

INVISIBLE ENVISIONING

Envisioning to solve a visible and familiar problem can be difficult, but not
as difficult as envisioning the invisible and unfamiliar. When the COVID-
19 (SarS-19) virus began its global movement in late 2019 and early 2020,
a researcher at the Center for Systems Science and Engineering (CSSE) at
Johns Hopkins University, Ensheng (Frank) Dong,[21] whose family lived in
China, near the location of the outbreak in Wuhan, designed a geospatially
enabled website, with the support of the Director of the CSSE, Dr. Lauren
Gardner, to track the global spread of this invisible monster.[22]

For part of 2020, the Johns Hopkins CSSE website became the most searched website on the globe. Its daily revisions, accurate to county, regional, and smaller administrative areas over the entire globe, had become the source of choice for many individuals and organizations who are trying to track, make sense of, or manage the risks associated with this unprecedented and unforeseen pandemic. Yet the global popularity of the site is not matched by an equal accuracy of the sources of information across the globe. Not every country has invested equally in its public health infrastructure, or GIS technology, or professional training and practices in both these fields. For political reasons, some countries or parts of countries chose to conceal their data.

Like Leonardo Da Vinci, who did not know the plans of Borgia's enemies, Frank Dong's envisioning capability had to deal with missing, incomplete, or possibly deceptive information. In many areas of the world, GIS professionals work assiduously every day to provide timely, accurate, and complete information to revise their local public health records and the Johns Hopkins University website. Yet in other parts of the world, the Central African Republic, for instance, no significant medical or public health infrastructure exists to track, much less record, changes in public health over a specified area. And that country is not unique in being beset by local conflicts, other severe medical and public health challenges, and a dearth of resources or government effectiveness.[23] In other parts of the world, evidence exists that the public health information supplied by local governments, even in wealthier countries, was distorted or deceptive. In Russia, evidence in 2020 indicated a substantially unlikely variance in the death rates year over year between 2018, 2019, and 2020.[24] Similarly, in Iran, the published health reporting about only a few COVID infections has been undercut by geospatial imagery showing the digging of mass graves.[25]

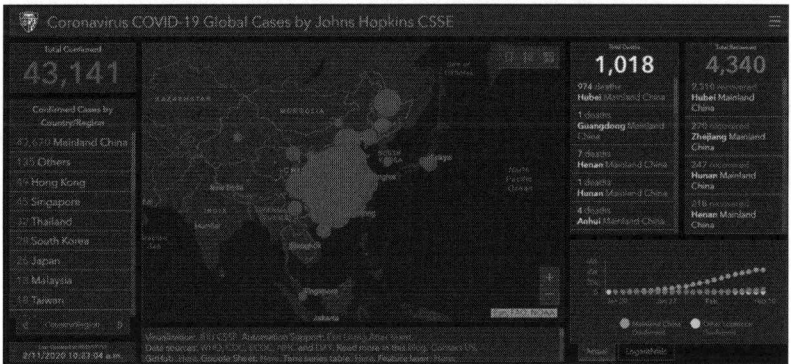

FIGURE 2.3   Screen capture of Johns Hopkins COVID-19 Dashboard.

Source: Coronavirus Resource Center Johns Hopkins University.

The Johns Hopkins COVID-19 site illustrates the potential of geospatial technologies and an enduring challenge for geospatial intelligence. The site's reliance on GIS technology, shared data formats, and volunteered spatially registered health information enables daily revisions on a global basis as well as local analyses based on the various political levels of public health infrastructure. Routinely, geospatial technology enables the tracking, comprehension, and discoveries of new outbreaks. Yet for all the value of modern digital GIS-based technology, a greater value comes from the humans who use it to assist their thinking to uncover deceptive, incomplete, or missing data. As the global vaccination programs emerge, GIS technology can measure the effects of various governmental health recovery plans. But the technology alone cannot identify where data are incomplete, as in regions with little identified public health infrastructure. Technology cannot identify where the data are misleading, as open source reporting from places as relatively affluent as St. Petersburg, Russia, Iran, and the US states of Florida and New York.[26] Geospatial analyses also indicate deceptive behaviors surrounding COVID, as shown by the commercial imagery analysis of Iranian graveyards.[27] Through noticing the discrepancies between the reporting and the accounting, geospatial analysts have seen and reported the implausible, and probably the unlikely, accounting of unexplainable changes in a global health pattern.

For as long as humans have congregated, there has been conflict, and for an equally long time, humans have sought to gain information to provide them an advantage by reducing uncertainty in future conflicts. This is one definition of intelligence.[28] And in the assessment of information, the envisioning, discovery, recording, comprehending, and tracking of "differences that make a difference," in Geoffrey Bateson's phrase,[29] humans frequently demonstrate their unique contributions to geospatial intelligence.

NOTES

1 Dreams and imaginative fantasies are also forms of seeing differently, but they are far less useful for assessing risks and measuring locations.
2 Strathern, Paul. *The Artist, The Philosopher, and the Warrior: Da Vinci, Machiavelli, and Borgia and the World They Shaped* (New York, NY: Random House, 2009), pp. 158–162.
3 *The Notebooks of Leonardo Da Vinci, Compiled and Edited from the Original Manuscripts by Jean Paul Richter in Two Volumes, Volume II* (New York, NY: Dover Publications, 1970), Plate CXI.
4 Issacson, Walter, *Leonardo Da Vinci* (New York, NY: Simon & Schuster, 2017), pp. 342–344.

5 My colleague, Robert T. Clark, points out that viewing the earth from aircraft or satellites presents "the ultimate high ground" *Geospatial Intelligence: and Origins* (Washington, DC: Georgetown UP, 2020), pp. 123–138.

6 Clark is also responsible for the first formulation of the drivers for what we now describe as geospatial intelligence. *Geospatial Intelligence*. Preface, p. xi.

Strathern, Paul. *The Artist, The Philosopher, and the Warrior: DaVinci, Machiavelli, and Borgia and the World They Shaped* (New York, NY: Random House, 2009), p. 160.

7 Issacson, *Leonardo Da Vinci*. pp. 200–203.

8 Wilford, John Noble. *The Mapmakers*, revised edition (New York, NY: Vintage, 2001), pp. 114–117.

9 Wilford, *The Mapmakers*, pp. 155–161.

10 Wilford, *The Mapmakers*, pp. 117–141.

11 Murphy, Justin D. *Military Aircraft: Origins to 1918*. (Santa Barbara, CA: ABC-Clio, 2005), p. 8.

12 Central Intelligence Agency. Intelligence in the Civil War (Washington, D.C.: Central Intelligence Agency, 2007), pp. 31–33; Glantz, Edward J. "Guide to Civil War Intelligence," *The Intelligencer: Journal of U.S. Intelligence Studies*, Volume 18, Number, 2, Winter/Spring 2011, p. 58; Murphy, Justin D. *Military Aircraft: Origins to 1918* (Santa Barbara, CA: ABC-Clio, 2005), pp. 10–11.

13 https://en.wikipedia.org/wiki/History_of_military_ballooning, viewed 3 May 2021.

14 https://en.wikipedia.org/wiki/Aerial_reconnaissance, viewed 3 May 2021.

15 Kennedy, Robert F. *Thirteen Days: A Memoir of the Cuban Missile Crisis*, with an afterward by Richard E. Neustadt and Graham T. Allison (NY: Norton, 1971). pp. 1–2.

16 Brugioni, Dino, A. *Eyeball to Eyeball: The Inside Story of the Cuban Missile Crisis* (New York, NY: Random House, 1990), pp 230–231.

17 Kilday, Bill. *Never Lost Again: The Google Mapping Revolution That Sparked New Industries and Augmented Our Reality* (New York, NY: HarperCollins, 2018), p. xv.

18 In-Q-Tel invested in Keyhole in February 2003. www.iqt.org/news/in-q-tel-announces-strategic-investment-in-keyhole/, viewed 24 March 2021; Kilday's book gives the date as March 2003, p. 79.

19 www.wsj.com/articles/SB109888284313557107, viewed 24 March 2021.

20 https://en.wikipedia.org/wiki/Coverage_of_Google_Street_View, viewed 24 March 2021.

21 Dong E, Du H, Gardner L. "An Interactive Web-Based Dashboard to Track Covid-19 in Real Time," *Lancet Inf Dis*. Vol. 20, No. 5,

pp. 533–534. doi: 10.1016/S1473-3099(20)30120-1, viewed 27 June 2021, https://coronavirus.jhu.edu/map.html.

22 Dr. Cassandra Hanson, of the Krieger School at Johns Hopkins University, first used this term to describe COVID-19.

23 www.aljazeera.com/indepth/features/central-african-republic-colos sal-struggle-covid-19-200421142222924.html, viewed 27 June 2021. https://en.wikipedia.org/wiki/Health_in_the_Central_African_Repub lic, viewed 27 June 2021.

24 Russia Leningrad information about COVID, www.theguardian. com/world/2020/jun/04/st-petersburg-death-tally-casts-doubt-on-russian-coronavirus-figures, viewed 27 June 2021, and national underreporting www.themoscowtimes.com/2020/07/10/new-figures-suggest-russias-coronavirus-death-toll-underreported-a70848, viewed 27 June 2021.

25 https://thearabweekly.com/iran-admits-under-reporting-coronavi rus-figures-it-faces-biggest-crisis-1979, viewed 27 June 2021; www. space.com/iran-coronavirus-graves-satellite-images.html, viewed 27 June 2021.

26 Florida Information. www.npr.org/sections/coronavirus-live-upda tes/2020/05/19/859119865/florida-ousts-top-covid-19-data-scient ist; New York. https://news.yahoo.com/ap-over-9-000-virus-224834 053.html.

27 https://coronavirus.jhu.edu/map.html, viewed 28 June 2021.

28 Clark, Robert M. *Intelligence Analysis: A Target-centric Approach* (Washington, DC: CQ Press/Sage, 2020), p. 22.

29 Bateson, Geoffrey, "Form, Substance, and Difference," in *ETC: A Review of General Semantics*, Vol .72, No.1 (January 2015), pp. 90–104. Published by the Institute of General Semantics, Stable URL: www. jstor.org/stable/24761998, Accessed 04-05-2021. This was the 19th Annual Korzybski Memorial Lecture, delivered 9 January 1970, under the auspices of the Institute of General Semantics, pp. 94–95.

BIBLIOGRAPHY

www.aljazeera.com/in-depth/features/central-african-republic-colossal-struggle-covid-19-200421142222924.html.

https://thearabweekly.com/iran-admits-under-reporting-coronavirus-figu res-it-faces-biggest -crisis-1979.

Bateson, Geoffrey. "Form, Substance, and Difference," *ETC: A Review of General Semantics*, Vol. 72, No. 1, January 2015, pp. 90–104.

Brugioni, Dino A. *Eyeball to Eyeball: The Inside Story of the Cuban Missile Crisis* (New York: Random House, 1990).

CIA. *Intelligence in the Civil War* (Washington, D.C. Central Intelligence Agency, 2007).

Clark, Robert M. *Geospatial Intelligence: Origins and Evolution* (Washington, D.C.: Georgetown UP, 2020).

Clark, Robert M. *Intelligence Analysis: A Target-centric Approach* (Washington, DC: CQ Press/Sage, 2020).

Dong E, Du H, and Gardner L. "An Interactive Web-Based Dashboard to Track COVID-19 in Real Time," *Lancet Inf Dis.* Vol. 20, No. 5, pp. 533–534. doi: 10.1016/S1473-3099(20)30120-1, viewed 27 June 2021, https://coronavirus.jhu.edu/map.html.

Glantz, Edward J. "Guide to Civil War Intelligence," *The Intelligencer: Journal of U.S. Intelligence Studies*, Vol. 18, No., 2. Winter/Spring 2011.

www.theguardian.com/world/2020/jun/09/st-petersburg-death-tally-casts-doubt-on-russian-coronavirus-figures

Google Earth. "Imola, Metropolitan City of Bologna, Italy." Imagery date 9/12/22 44°21'11.72"N 11°41'32.08"E.

Howell, Elizabeth "Satellite Images Show Iran's Mass Graves for Coronavirus Victims," last updated 13 March 2020, www.space.com/iran-coronavirus-graves-satellite-images.html.

www.iqt.org/news/in-q-tel-announces-strategic-investment-in-keyhole.

Isaacson, Walter. *Leonardo Da Vinci* (New York, NY: Simon and Schuster, 2017).

Kennedy, Robert F. *Thirteen Days: A Memoir of the Cuban Missile Crisis.* with an afterward by Richard E Neustadt and Graham T. Allison (New York, NY: Norton, 1971).

Kilday, Bill. *Never Lost Again: The Google Mapping Revolution That Sparked New Industries and Augmented Our Reality* (New York, NY: Harper Collins, 2018).

www.themoscowtimes.com/2020/07/10/new-figures-suggest-russias-coronavirus-death-toll-underreported-a70848, viewed 27 June 2021.

Murphy, Justin D. *Military Aircraft: Origins to 1918* (Santa Barbara, CA: ABC-Clio, 2005).

www.npr.org/sections/coronavirus-live-updates/2020/05/19/859119865/florida-ousts-top-covid-19-data-scientist.

NRO. Center for the Study of National Reconnaissance. *The Era Before KENNEN/KH-11.* www.nro.gov/Portals/65/documents/foia/declass/HISTORICALLY%20SIGNIFICANT%20DOCs/NRO%2060th%20Anniversary%20Docs/SC-2021-00002_C05097836.pdf.

Richter, Jean Paul (ed.). *The Notebooks of Leonardo da Vinci,* Compiled and Edited from the Original Manuscripts in Two Volumes. Volume II (New York: Dover Publications, 1970). www.rct.uk/collection/912284/a-map-of-imola.

Strathern, Paul. *The Artist, the Philosopher, and the Warrior: Da Vinci, Machiavelli, and Borgia and the World They Shaped* (New York, NY: Random House, 2009).

Wilford, John Noble. *The Mapmakers*, revised edition (New York, NY: Vintage, 2001).

https://en.Wikipedia.org/wiki/AerialPhotography.

https://em.wikipedia.org/wiki/Aerial Reconnaissance.

https://en.wikipedia.org/wiki/Coverage_of_google_street_view.

https://en.wikipedia.org/wiki/health-in-the-central-african-republic.

www.wsj.com/articles/SB109888284313557107, viewed 24 March 2021.

https://news.yahoo.com/ap-over-9-000-virus-224834053.html.

CHAPTER 3

Discovery

INTRODUCTION

Discovery is the second intellectual activity in geospatial intelligence, and chance and risk make it different from the other four activities. While the other four intellectual activities—envisioning, recording, comprehending, and tracking—can be expressed with an active word—discovery does not always result from prior effort. Chance, risk, opportunity, deception,, and uncertainty surround and sometimes preclude discovery. Sometimes discovery is influenced by probability, sometimes it is influenced by luck, and sometimes, in spite of preparation and thoughtful action, it just does not occur. The history of intelligence, and the history of geospatial intelligence, is full of unmade discoveries, found only in retrospect. But the successful discoveries improved mapmaking, plotted previously unseen relationships, found intelligence in aerial photographic images, and succeeded in changing historical outcomes.

John Snow's analysis of the outbreak of Cholera and the Broad Street pump in London in 1854 shows that spatial discovery can result from plotting or arranging known information on a map or chart. Snow's map helped to make a previously unknown relation between a contaminated water source and an incurable disease clear, and provided him with a key to effective action in his campaign to teach and influence others in medicine to change their thinking about the source of cholera.

The unknown Italian intelligence officer and aviator who took the first known aerial combat photographs in the Italo-Turkish war of 1912 made a technological discovery that remains in use today. A few years later his initiative was refined and duplicated tens of thousands of times in World War I, when intelligence officers combined aerial photography with the 19th century discovery of stereo photography to change terrain analysis forever.

Many additional discoveries from aerial intelligence occurred in World War II, and that war also had several examples of failures to

DOI: 10.4324/9781003436836-4

discover. In the case of the British torpedo attack on the Italian fleet at Taranto in 1940, the discoveries that would change the course of the war were made by the British and the Japanese and the failures to discover were made by the Italians and the Americans.

In 1943, a photographic image from a diverted mission allowed British photointerpreters to fortuitously discover a number of German strategic missile programs at Peenemunde. Even though this observation would lead to large air raids with high casualties, the Allies could not prevent the German deployment of these weapons or interfere effectively with their operation. German awareness of the British discovery of Peenemunde caused them to create countermeasures that made future British and Allied discoveries of these weapons more difficult.

In 1978, 34 years after an initial observation over Poland on aerial photography, a retrospective analysis of the aerial photography of Auschwitz illustrates a failure to discover at the time of the photographic collection. The analysis of the Auschwitz imagery also shows how geospatial intelligence can be limited without additional context and other sources of information.

DISCOVERY

Leonardo da Vinci embodied the parallel tracks of geospatial thought and geospatial technology, and since his time, they have been rarely combined in any single human being. For three centuries after Da Vinci envisioned the world differently, the technology of map making would proceed from hand-drawn experimentation through calculation to duplication on a printing press, and incrementally, through increasing use of mathematics, observation, and surveying, toward defining locations ever more precisely. Throughout this long interval, nation states arose and sought to define their borders, European nations began to explore other continents seeking wealth and territory, and mapmakers in many countries began to capture their discoveries accurately and to create mathematical cartography. Armies sought to characterize terrain accurately before combat. Sailors in navies and merchant fleets sought safe passages at sea, and many countries and some colonies worked to define their borders more precisely. As a byproduct of global exploration, individuals made conceptual, technological, and industrial discoveries to create terrestrial and maritime knowledge. That knowledge provided advantages for the discoverers in their decision-making. From the 16th to the 19th centuries, mapping became more precise, and nations created national mapping organizations to make their discoveries politically and militarily useful.

Improved tools for making mathematical measurements made the maps and charts more precise. In the 16th century, Jean Frenel in France combined an odometer, much like Leonardo's hodometer, with a quadrant to ascertain his location and from that to measure the length of a degree of latitude.[1] Surveying tools improved with the invention of better optics. In the early 18th century, the addition of telescopes to theodolites allowed for more accurate measurements of elevation as well as location and distance. Throughout the 19th century European scientists and explorers made nearly unimaginable journeys to take measurements in remote and unexplored locations to improve the accuracy of their measurements across wider areas of the globe. In France the Cassinis and in England William Roy and Jesse Rasmden engaged jointly on a Cross-channel triangulation to measure precisely the distances between the two countries. While these efforts were scientific in nature, they influenced the thinking of the military leadership of both countries. The Carte de Cassini became the basis for French topographic work, and the efforts of William Roy led George III to create the Ordnance Survey in England in 1781.[2]

Efforts continued on every continent throughout the first half of the 19th century to extend precise measurements over more of the planet. The technologies of mathematical cartography proceeded from experimentation to industrialization. These technologies resulted in far more accurate maps and charts that improved safety and reliability for travelers. The industrialization of railroads, telegraphy, and transoceanic shipping led to global standards for latitude and longitude, as well as the creation of regulated time zones, all of which increased geographic discoveries.

Yet, all these efforts to increase the accuracy of mapping the parts of the world did not increase the human capability to make more discoveries about the world. In the middle decade of the 19th century two discoveries occurred: one revealed the power of location in identifying the cause of an invisible medical problem, and the other introduced a new way for man to look at and record information about the planet.

MAPPING THE INVISIBLE

John Snow was the first man to map the invisible, and he changed how people think about proximity and the sources of disease.[3] As Da Vinci changed how humans think about the physical relations among locations, Snow's discovery would change how humans think about their relation to their immediate environment. Snow accomplished this with only his eyes, his ability to think spatially, and his medical training. In 1854, John Snow was attempting to isolate the cause of a cholera epidemic in London. At that time, there was no current effective treatment for cholera. Snow had been skeptical about the prevailing theories which held that the cause of the disease was airborne.

Snow's insight that drinking contaminated water might transmit cholera led him to examine the locations of the victims and their water sources. His discovery of the spatial relation between the continuing contagion, the infected victims, and a contaminated water pump was the first instance of mapping the invisible. Snow's challenge was how to relate the residences of the infected humans to the map of London to see if he could identify a pattern from his evidence. His data gathering among the victims, at considerable personal risk, led him to observe that different locations in the neighborhood in which the disease was raging drew their water from different sources. On the basis of his initial findings, Snow had the Broad Street pump handle removed. His action and the subsequent drop

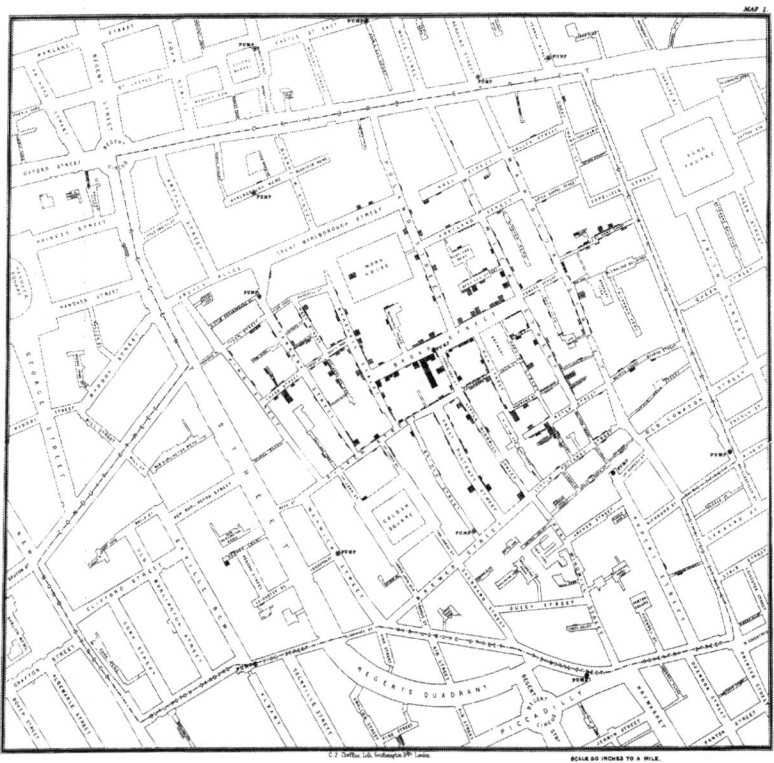

FIGURE 3.1   John Snow's map of the Broad Street Pump, London.

Source: Public domain.

in the number of cholera cases gave him the initial evidence to support his theory that cholera was waterborne.

When Snow had the pump handle removed, he was able to bound the extent of the infection problem. His discovery about spatial boundaries enabled him to reach a unique insight. Snow's thinking about the locations of the infections supported the line of thought that eventually changed medical thinking about the origins of cholera. Snow's map influenced others in the medical community to begin to think differently. He also began a line of intellectual inquiry that would enable the 21st century work of Frank Dong on COVID-19 at Johns Hopkins. Snow's map convinced others that contaminated water rather than airborne contagion caused the cholera outbreak in London (Figure 3.1).

Until medical microscopy and geospatial technology could catch up with his spatial thinking, the insight Snow discovered about the relation of humans, proximity, and disease would not be improved on.

A PERMANENT BIRD'S-EYE VIEW

A few years after Snow's map of part of London, in North America, James Wallace Black took a photograph from a hot-air balloon of part of Boston. It would turn out to be the first preserved aerial photograph (Figure 3.2).

Previously aerial photographs had been taken from balloons in France but none survived.[4] After Black's success, Thaddeus Lowe attempted to apply balloon photography to military intelligence and observation in the US Civil War (1860–1865). In spite of the resources and impetus supplied by the American Civil War, the new technology of aerial photography would not be sufficiently developed to make a difference in that war.

Throughout the latter half of the 19th century, cartography improved and experiments continued in aerial photography in the US and Europe, involving kites, balloons, and even pigeons. Other 19th century technologies, namely railroads and telegraphy, radically changed how military forces operated and increased the demands for accurate timekeeping and precision location during the US Civil War (1860–1865).[5] It would take the War to end all Wars, started in 1914 by competing European nations, to provide the impetus to combine military mapping, Da Vinci's envisioning, and aerial photography into a new kind of military intelligence.

Aerial photography in World War I turned Da Vinci's initial insight at Imola into a matter of military routine for engineer and intelligence officers on both sides. The aerial photograph, first taken in France in the 1850s, first recorded in Boston in 1860, became a routine daily activity

Balloon View of Boston Taken October 13 1860
By J.W. Black

FIGURE 3.2   James Wallace Black's aerial photograph of the north end of Boston.

Source: The Metropolitan Museum of Art, New York, Gilman Collection, Purchase, Ann Tenenbaum and Thomas H. Lee Gift, 2005.

for the principal combatant forces. From a 1912 experiment in the Italo-Turkish War, by 1915 in France and Belgium military photography from French and British aircraft had become another customary operational practice.[6]

In the three years after an Italian pilot proved the value of military aerial photography, it had become an effective planning tool for ground combat. A 1915 British and French attack in the First World War, Neuve Chapelle, south of Ypres in Belgium and north of the Somme battlefield is little known and less remembered. But it marked the first time in the history of combat that interpretation of aerial photography shaped future battle planning.

The fighting from 10–13 March 1915 at Neuve Chapelle was not militarily decisive. Thankfully it did not achieve the scale of slaughter of more famous World War I battles. But Neuve Chapelle marked the first time that military planners used aerial photography to shape future military operations.[7] A British officer succeeded in showing his superiors how aerial photography could help the planning of offensive and defensive general operations. While the British and French armies achieved only some of their objectives, the forces were able to deconflict their axes of advance and achieve early successes on the basis of unique intelligence gained from aerial photographs.[8]

After Neuve Chapelle, photointerpretation became more frequent and valuable in the continuing development of World War I military intelligence. In each of the major Allied forces, France, England, and the US, individuals—Eugene Pepin and Paul-Louis Weiller among the French, C.D.M Campbell and J. T. C. Moore-Brabazon among the British, and Edward Steichen and Billy Mitchell among the Americans—all increased the utility of aerial photography and wrote about the unique value of aerial photointerpretation.[9] Yet as the advocates from every nation noted, to make discoveries, the photointerpreter would have to deal with intrinsic demands and extrinsic deficiencies.

THE TRAITS OF A NEW MILITARY PROFESSION

Among the intrinsic demands on photointerpreters was the ability to see as a ground soldier as well as an artist. World War I practitioners and commentators on aerial photointerpretation wrote about the need for interpreters to be able to look at a photograph from an unfamiliar point of view while being able to distinguish man-made features from terrain features.[10] The interpreters also needed technical proficiency, as they had to identify artifacts on the photography introduced by chemical processing so that they would not confuse these processing artifacts with terrain or man-made features. And finally, the interpreters needed acuity, patience, memory, and curiosity.

Acuity was needed to notice on a photograph new terrain features and man-made effects, such as trenching or artillery damage. Patience was essential as to be certain of their findings, the interpreters had to go over and scour dozens of square miles of terrain at varying resolution. After interpreters had looked at hundreds of photographs, memory became as essential as the first two attributes. Photointerpreters had to remember what each scene looked like on the prior observation. If they observed new features, they also had to remember if they had seen anything like the new observations, and where and when they had seen them. All these instances of prior observation could be brought to bear successfully to defeat the intent of camouflage, concealment, or deceptive measures by the enemy. But in the absence of technical assistance, the memory of the photointerpreters held a great part of the context against which new photographs were evaluated. The pace of work and the limits of recording technology meant that many times the only repository for a large number of their prior observations was the storage device between the interpreters' ears.

Curiosity was essential. More than a century later, it remains a difficult challenge to measure and assess. In a high-volume, stressful, repetitious process, like wartime photointerpretation, it can be a difficult human attribute to sustain. In some instances, curiosity is noting the new, incongruous, or mis-located. In others, it is noticing the absence of change or expected observations, and in yet other cases, it involves repeatedly examining the incomprehensible until it makes sense or partial sense. This trait was found in the best photointerpreters and imagery analysts. It is found in the best geospatial analysts.

And to be successful, photointerpreters required patience. Their work could be complicated by clouds, smoke, fog, missions that never returned, technical problems with the camera, film plates, focus, or bad processing in the lab. If the human characteristics of memory, acuity, patience, and curiosity were not brought to bear by the interpreters, enemy deception, camouflage, concealment, or denial could be successful.

The interpreters learned that stereoscopic photographs would provide additional information, and they also learned that some among their military audiences were unable to distinguish detail or to see in stereo. Stereo photography—two photographic images of the same area with at least 60% overlap—made terrain analysis possible over an area of denied territory. But stereo photography required either two aircraft flying in straight and level formation or the same aircraft to overfly an area twice. Overlapping photographs, viewed through a stereoscope, displayed the underlying terrain in three dimensions, which made possible the discovery of subterranean emplacements, trenches, hidden ground, and emplaced artillery. Yet, the requirement for stereoscopic photographs of the same terrain increased the risks and shortened the lives of reconnaissance pilots.

The intelligence value of photo observation in World War I was evident by the steady increase in the demands made on its creators. The creation of information from aerial photographs was followed quickly by the industrialization of the photographic, cartographic, and intelligence processes on the ground. All World War I aerial intelligence was hand-made, and in many commands the volume of requested information required that this work be done on a 24-hour basis.

While many hundreds, and at times, many thousands of photographs were duplicated, all this effort had to be translated every day for senior officers in the military audience that, like Bobby Kennedy many years later, could not see details on the photographs. This translation of information from photography was done in all the Allied forces. Restitution, as the process was then called, converted photo-derived intelligence into inked markings on paper maps. This additional time-consuming step was necessary as at that time European senior military leaders had learned to see the battlefield terrain on maps, and many of them could not use aerial photography to envision changes in the battle lines or terrain features. By the third year of World War I, restitution was done daily for the senior military leaders of all the armies. The challenges of getting these leaders to see the results of new technologies seem to be an enduring legacy in the history of geospatial intelligence and its antecedent disciplines. Fortunately for the Allied forces, and less fortunately for the Germans, individual younger officers of lower ranks in all Allied armies could envision the value and the future of aerial photography.[11]

After the horrific casualties to both sides of Verdun and the Somme offensive in 1916, all combatants sought to use photo-intelligence and every other form of intelligence to reduce the human casualties. The French and British armies had reorganized their staffs to increase the number of intelligence officers, particularly photo-intelligence officers. By 1916, both alliances and their air forces had recognized the need to improve aircraft speed, to provide fighter coverage on reconnaissance missions, and to improve cameras to diminish the amount of work that had to be done in the aircraft. Greater resources on the ground complemented improvements in the air. The number of mobile photo labs, men, and equipment increased. The Allied photographic processing improved so that under optimal conditions aerial photography could be delivered to headquarters within one hour after the aircraft had landed.

By 1918 all major combatant countries had integrated aerial reconnaissance into their daily operational military planning. Aerial photography had supplemented aerial observation as the measure of determining effectiveness for artillery fire. As in 1918 the addition of US forces and materiel began to shift the tide of battle in favor of the Allies, and the use of aerial reconnaissance continued to grow. But the November 1918 Armistice ended the growth and development in military aerial reconnaissance.

INTERWAR DISCOVERIES

In the interwar period, developments in aerial photography continued, but they were isolated and not introduced systemically into any military force. In the US Sherman Fairchild continued to develop dedicated aerial cameras and eventually dedicated aircraft for taking photographs at higher altitudes.[12] Fairchild, who would develop many aviation inventions, created the first profitable aerial survey business in the Western Hemisphere. But in the early 1930s the worldwide depression, with its global economic impact, constrained the markets for aerial photography.

Between the World Wars, the European and American military did little innovative work with aerial photography. The British flew mapping surveys over parts of their empire in Egypt, Iran, and India[13] but they did little aerial intelligence collection until the late 1930s. In the US a few individuals sustained aerial experimentation, notably Sherman Fairchild's civilian work with camera development, and George W. Goddard's Army Air Forces experiments with aerial night photography.[14] But in the middle of the 1930s, the changing politics in Europe resulted in two new developments in aerial reconnaissance.

As Germany began to rearm its military after 1935, it began clandestine aerial surveillance from commercial Lufthansa flights over Great Britain and France to discover more about their military and industrial preparedness. The British, equally concerned about the German military buildup, enlisted an Australian businessman, Sidney Cotton, in their reconnaissance efforts. Cotton, a globally successful businessman and aviation enthusiast, began to fly reconnaissance missions over Libya and Germany for the British government in his privately owned Lockheed Electra.[15] On these missions Cotton discovered that altitude and airspeed mattered more than ever to pilot safety and good aerial photography. As a result of Cotton's pre-war efforts, the British were able to discover reliable intelligence about the location of many of the German Navy's surface vessel and submarine construction yards at the onset of World War II.

In September 1939 the start of war in Europe spotlighted the intelligence weaknesses of each combatant country, and how much each would have to relearn about photointerpretation. The German military efforts at aerial reconnaissance relied on high-altitude aircraft and did not change throughout the war.[16] The British, initially less prepared than the Germans, after taking many casualties on unsuccessful missions, recognized that slow converted bombers were ill-suited for reconnaissance and shifted to fast, unarmored fighter aircraft capable of both high-altitude and high-speed reconnaissance.

In the early years of World War II, the UK and the US had to relearn most of the aerial reconnaissance lessons of World War I. The English,

led by the pre-war clandestine work of the Australian Sidney Cotton
and F. W. Winterbotham, and facing manpower shortages from the onset
of the war, incorporated women as photointerpreters beginning in June
1940.[17] British women interpreters made some of the most important
discoveries of the war. After the 7 December 1941 Japanese attack on
Pearl Harbor, the US had a similar relearning curve, but profited from
the British experience which they generously shared with the US Navy
and the US Army Air Corps. The US military also trained and employed
women photointerpreters.

In May 1944, After the US invasion of Africa and the Allied invasions
of Sicily and Italy, the US and the UK combined their European photo-
reconnaissance efforts into the Allied Central Interpretation Unit (ACIU)
at Medmenham, near Oxford. As in World War I, the scale of the
European photo-intelligence efforts had grown rapidly. The ACIU by late
in 1944 grew to more than 1,700 personnel.[18] Three different missions—
strategic infrastructure analysis, invasion preparations, and the search for
German strategic weapons—drove the volume of the ACIU work past
any imagined capability. The analysis of German strategic infrastructure comprised two intelligence tasks—discovering the locations of the
research, development, and test ranges for strategic weapons, and subsequently developing and comprehending target information to support
mass bombing raids, and bomb damage assessment after these raids.

The British developed a new tradecraft for photointerpretation that
did not exist in World War I. It would shape the future of geospatial
intelligence in the Cold War. This was military Research & Development
(R&D) analysis. This tradecraft grew out of their pre-war intelligence
effort to determine the rate of German air force production. During the
war, the Royal Air Force flew reconnaissance missions against known aircraft production sites, and against suspected R&D sites.

In the event of cloud cover, on any particular day, aircraft reconnaissance missions routinely had backup targets. In the spring of 1942,
a reconnaissance mission planned to photograph Berlin found the city
cloud-covered, and the pilot continued northward to the German Baltic
seacoast. He photographed an area near Peenemunde on the northeastern
German coast. On 15 May 1942 the interpreters of CIU in Medmenham
saw a number of unusual objects that they could not identify.[19]

At test facilities photointerpretation generally requires multiple
images to discover prototype weapons systems. As in World War I,
photointerpreters in World War II often had to rely on manually derived
information about target locations. Much preliminary analysis was
accomplished using map/film correlation and in-scene scaling to determine relative distances.[20] And for all the photographs acquired, the
time over target, the exact elevation of the aircraft, and the azimuth of
approach were all manually calculated, sometimes under extreme stress.

Peenemunde was a German R&D site for aircraft, rockets, and missiles. On first examination, the British photointerpreters saw numerous objects and facilities that they could not identify. Subsequent study and information from other intelligence sources identified aircraft without propellers and cylindrical, torpedo-shaped objects later identified as missiles. The discoveries so concerned the British government that it attacked Peenemunde in August 1943 with a 500-bomber raid. While the British bombed this facility with the hope of destroying its infrastructure and killing the scientists and specialists who lived there, it failed to achieve either of these objectives. The weapons first observed at Peenemunde, the V-1 cruise missile and the V-2 ballistic missile would in 1944 continue the attacks on civilian targets over England. They would also be used in 1944 and 1945 to attack Dutch cities until the end of the war in spite of the Allied advances in Europe. The Allies mounted an intensive aerial reconnaissance in the spring of 1944 to find V-1 and V-2 sites, but the search was only partially effective.

German V-weapons began to strike England in June 1944, shortly after the Normandy Invasion.[21] While militarily ineffective, these missiles effectively terrorized the British civilian population and leadership as there was no warning of their approach or impact area. The discovery of their deployed launch sites was a far more difficult challenge for the photointerpreters than finding their permanent sites. Finding the V-1 launch sites started a photographic area search that required hundreds of aircraft missions to be flown over western France, Belgium, and Holland. The Germans began to construct fixed launch positions for the V-1, much like the ones identified at Peenemunde, but Allied success at discovering and bombing these caused the Germans to adopt camouflaged temporary and mobile launch sites for their V-1 deployments into France and Belgium. The German V-1 and V-2 deployments created the first tracking problem for photographic intelligence, and it occurred at the same time as the late stages of the Allied preparations for the Normandy invasion. The confluence of these priority intelligence requirements greatly increased the work for the British and US analysts and support staff in the ACIS at Medmenham, and the ACIS had to increase its processing and duplication capabilities.[22]

WARNING FROM DISCOVERY

World War II brought about new discoveries from aerial photography, and strategic warning was one of them. In the late spring and summer of 1940, after the Dunkirk evacuation, England remained the only European nation opposing Germany, and the invasion of Britain seemed imminent. German bombing of England began in that summer of 1940.

Yet throughout, the British photointerpreters were able to continue to monitor the German logistical buildup for their intended invasion.

From their previous two military cross-channel deployments into France and Belgium in 1914 and 1940, the British were fully aware of the amount of naval transport necessary to move a military force across the English Channel, and more importantly, the amount of naval transport, infrastructure, logistics, and protective forces it would take to sustain such a military invasion. This awareness, in conjunction with the Royal Air Force's frequent aerial observation of the Channel ports in France, Belgium, and Holland, provided the British government a daily status of how much German naval transport was available and what was lacking for a potential invasion of England from the continent.

The invasion threat diminished in October 1940, and BritIsh photointerpreters focused on warning about the two most serious strategic threats—German submarine construction and the infrastructure of the German Air Force. Britain depended on sea-borne replenishment for food, weapons, men from Commonwealth and other nations, and raw materials, so looking at German submarine production and operating bases was a priority. The Central Interpretation Unit focused equally on German aircraft production and testing. As Britain had been under continual nightly air attack from June 1940 through March 1941, these two issues commanded the attention of a significant number of British photointerpreters at their Central Interpretation Unit at Medmenham.

Yet, the discovery that may have had the most effect on World War II did not happen at Medmenham. It had taken place at a British photointerpretation unit in the Mediterranean in November 1940. British photointerpreters and daring low-altitude reconnaissance were central to the success of a British torpedo attack launched from an aircraft carrier on Taranto, the home port of the Italian fleet, in November 1940. While the Italians thought that their defensive measures were sufficient, Admiral Cunningham planned a carrier-based torpedo attack on the basis of prior-day, low-level aerial reconnaissance. The attack plan relied totally on the photo-reconnaissance, and the successful attack eliminated the Italian surface fleet from influencing the remainder of the sea war in the Mediterranean.

Even today, more than 80 years later, the attack on Taranto has implications for intelligence analysts. Gary Klein, in a brilliant comparison,[23] points out how the Japanese Naval leadership and the American Naval leadership looked at the same evidence after the British attack at Taranto and arrived at opposite conclusions about the potential feasibility of an aerial attack on the American Naval base at Pearl Harbor.[24]

The World War II use of photointerpretation for mission and target planning duplicated the World War I effort, but at a greater scale. The first recorded effort at using photointerpretation to assist in the planning for a seaborne invasion was a January 1940 experiment in California by General Mark Clark in the US Third Infantry Division exercise at Fort Ord. Reportedly, prior to the exercise, Clark paid a civilian pilot from his own pocket for the aerial collection.[25] Four years later, Clark's proof-of-concept bore fruit in planning for the Normandy invasion. To obtain greater detail, invasion reconnaissance missions were frequently flown at lower altitudes. Yet, aerial reconnaissance, while it could identify some obstacles, could not discern underwater obstacles. Some obstacles behind the beachheads were effectively camouflaged and some of the terrain analysis, particularly in the bocage or rural hedgerow area of Normandy and Brittany, failed to discern the military challenges the terrain would create for the expeditionary force. These failures of discovery proved serious for the invasion and the breakout from Normandy. Repeatedly, the allies flew reconnaissance missions of the actual and dummy invasion sites, and they also flew inland penetration missions over the French railroad system to identify chokepoints, such as bridges, tunnels, water towers, and railyards. By spring 1944, the Allied intelligence preparations for the Normandy invasion had grown to more than a million copies of photomaps of the invasion beaches, as well as more than a hundred newly created three-dimensional models of the same areas.[26]

AUSCHWITZ: DISCOVERING WHAT CANNOT BE ENVISIONED

But for all the World War II intelligence discoveries and successes, there was important intelligence that went undiscovered. No record exists of any attempt to use photointerpretation to discover the extensive German infrastructure that supported the Nazi program to exterminate the Jewish populations of Germany and the Eastern European countries occupied by the German armies. The German extermination program, known as "the final solution," was systemically planned, well supported by the German and European railway systems, with an infrastructure extensive enough in Germany to be mapped by the US Office of Strategic Services in mid-1944.

This June 1944 Office of Strategic Services (OSS) map displays the locations and approximate number of prisoners in the Nazi concentration camps throughout Germany. It shows the distribution throughout Germany of these camps for political prisoners, social undesirables, and Jews. Émigré, refugee, and neutral-country reporting collected by the Allies provided the information from which this map was created. While by June 1944 the general locations of the concentration camp infrastructure inside Germany were known, no systemic study has been undertaken

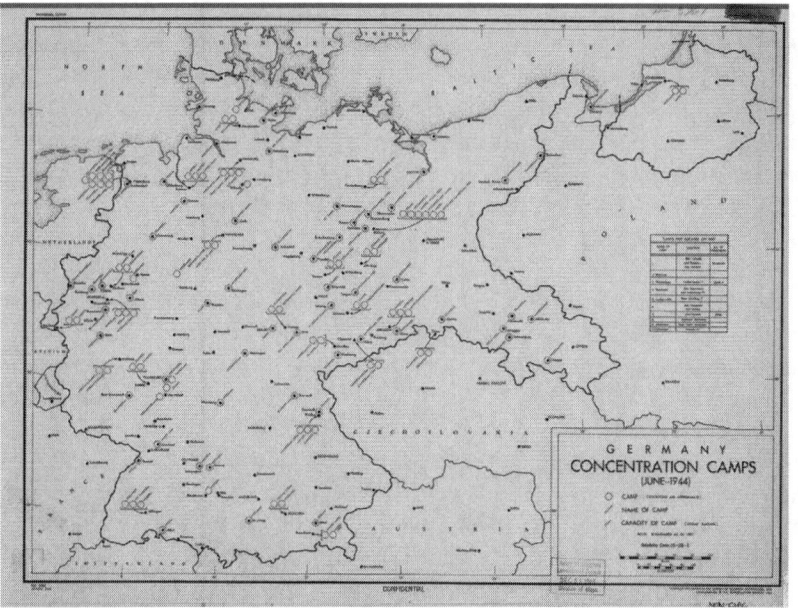

FIGURE 3.3   OSS 1944 Map of concentration camps in Germany.

Source: United States. Office of Strategic Services. Research and Analysis Branch. "Germany, Concentration Camps (June–1944)." Map. Washington, D.C.: OSS, 1944. Norman B. Leventhal Map & Education Center, https://collections.leventhalmap.org/search/commonwealth:ht25 07838 (accessed 3 May 2023).

to determine how much of the concentration camp infrastructure inside Germany had been photographed by the Allied reconnaissance effort (Figure 3.3).[27]

Far less information had been discovered during the war about the Nazi concentration camp infrastructure in eastern Europe. In 1978 a study was done of the Allied reconnaissance photo collection over the Nazi concentration camp in Poland at Auschwitz-Birkenau.[28] The retrospective analysis points out the limitations of looking at images with no other contemporary intelligence sources. The authors point out that all reconnaissance imagery had been taken with the intent of identifying military targets, and the nearby I.G. Farben factory for synthetic rubber and liquid fuels would have had far greater intelligence interest than the adjacent camps at Auschwitz.[29]

The Allies took nine photographs of this area between 4 April 1944 and 21 January 1945, the last day of the German evacuation of the Auschwitz camp, but all the aerial photography and the photointerpretation focused on the nearby I.G. Farben factory complex.[30] The photointerpreters, lacking any other sources of information, would have been able, according to the study, to identify the barracks, the extensive rail siding network, the security areas, and the separately secured areas that held buildings with roof vents, smokestacks, and nearby pits, but from aerial photography alone they would not have been able to discover the unimaginable events that took place there.

After World War II, as had happened after World War I, the Allies demobilized most of their military forces and scrapped most of their intelligence collection capabilities. Most of the Allied aerial photographic interpretation capability disappeared. At the surprise outbreak of the Korean war in 1950, the US Navy and Air Force had to recreate much of its military photointerpretation capability. The US discovery that the Soviet Union had nuclear weapons in 1948 and a means to deliver a number of them in 1953 led the CIA to create a Photo-Interpretation Division in the early 1950s. The need to discover what military capability was behind the Iron Curtain drove much experimentation, even with 19th century balloon technology.[31] By May 1956, the US had an aircraft that could fly deep into the USSR to discover and photograph strategic targets. Yet not all Cold War discoveries led to analytic success. A significant failed discovery of a Soviet Strategic facility occurred near the end of the Cold War. The delay in discovery was an outgrowth of the volume of the US imagery collection of the Soviet Union.

KRASNOYARSK

In the early 1980s, after more than 25 years of increasing the photographing and imaging[32] of the Soviet Union, an American satellite imaged a large construction project in the eastern Soviet Union, near the town of Krasnoyarsk. While the facility had been collected and photographed, that particular photograph was not analyzed. By this time, the US had discovered thousands of Soviet military and strategic targets. But the volume of imagery collection in the 1980s far exceeded the US capacity to look at all of it. This particular construction project, more than 1,000 miles from a Soviet border, was a large phased-array radar designed to track missiles. The remote location was one of the reasons that the image was not looked at. After its delayed discovery, in the judgment of the US arms control treaty experts, the location of the future radar constituted a treaty violation. The Soviets protested that they had built the radar at Krasnoyarsk so far from the coastline to offset the very high cost of building such an installation near the northern periphery of the eastern

Soviet Union.[33] But the out-of-the-way location of the radar meant that it was being constructed in a region where the US would consider it to be a treaty violation. More importantly, the out-of-the-way location also meant that the construction occurred in a remote region that was not a priority for US imagery analysts to search. They did not discover the construction until 1983, two or three years after it had begun.

While the political consternation about Krasnoyarsk focused on the potential treaty violation, the intelligence consternation grew out of the realization that between 1972, when the KH-9 went on orbit, and 1983 the US had collected so much aerial photography and imagery that it hadn't been able to look at all of it. The US Intelligence Community missed the construction at Krasnoyarsk for two or three years.[34] By 1982 or 1983, the volume of imagery from the KH-8 and KH-9 film-return and the KH-11 digital satellites exceeded the intelligence community's capability to search it all. The failure to discover Krasnoyarsk, ultimately not a serious strategic threat to the US, is noteworthy. It points to the risk to future discoveries caused by the growth in the volume of imagery collection capabilities exceeding the available attention of those charged with looking at the imagery. The chapter on recording discusses the implications of this growth in imagery data.

NOTES

1 Wilford, *The Mapmakers*. pp. 112–113.
2 Wilford, *The Mapmakers*, pp. 139–150; The website for the Ordnance Survey has a useful history. www.ordnancesurvey.co.uk/about/history.
3 An analysis of John Snow's creative insight regarding the Broad Street Pump is in Gary Klein's *Seeing What Others Don't: The Remarkable Ways We Gain Insights* (New York, NY: PublicAffairs, 2013), pp. 69–74. For a longer treatment, see Steven Johnson's *The Ghost Map: The Story of London's Most Terrifying Epidemic, and How It Changed Science, Cities, and the Modern World* (New York, NY: Penguin, 2008). www.gislounge.com/john-snows-cholera-map-gis-data/#:~:text=John%20Snow's%20well%20known%20cholera,epidemic%20to%20show%20disease%20clusters.
4 Nadar, Felix, Gaspard Felix Tournachon and Thomas Repensek. "My Life as a Photographer," in *October*, 1978, No. 5, Photography (Summer 1978), pp. 6–28. Nadar describes his difficulties in attempting aerial photography on page 24. Eder dates Nadar's attempt to 1858. *History of Photography*, pp. 393–394.
5 Gallison, Peter. *Einstein's Clocks, Poincare's Maps: Empires of Time* (New York, NY: Norton, 2003).

6 Finnegan, Col Terrence J. *Shooting the Front: Allied Aerial Reconnaissance and Photographic Interpretation on the Western Front—World War I* (Washington, D.C.: NDIC Press, 2006). pp. 47–51, Finnegan's book is a masterful and detailed history of the Allied photo-intelligence efforts, accomplishments, technologies, training, and issues.

7 Finnegan, *Shooting the Front*. pp. 39–42, 44–47.

8 Finnegan. p. 47.

9 Finnegan, Pepin, pp. 198–199, 226; Weiller, pp. 58, 60–61, 119–120. Campbell, pp. 45–47, and Moore-Brabizon, pp. 72, 84; Steichen, pp. 477–480, Mitchell, pp. 229–230.

10 Finnegan, pp. 445–457.

11 Finnegan, *Shooting the Front*. Mitchell, pp. 124, 224, 229–230; Pepin, pp. 198–201; Trenchard, pp. 45–46.

12 Burtch, R. Class Notes Sure 340, 2009, *History of Photogrammetry*, Ferris State University. pp. 21–22.

13 Conyers, Roy Nesbit. *Eyes of the RAF: A History of Photo-Reconnaissance* (Washington, D.C.: Sutton, 1996), pp. 54–58.

14 Goddard, George W. *Overview: A Life-Long Adventure in Aerial Photography* (Garden City, NJ: Doubleday, 1969), pp. 127–142.

15 Constance, Babington-Smith, *Evidence in Camera*, pp. 5–10; Lloyd, Steve. "Cotton's Covert Cameras: New Perspectives via Reverse Engineering," *Medmenham Magazine*, Spring 2020, pp. 22–25, https://en.wikipedia.org/wiki/Sidney_Cotton,viewed 4 May 2021.

16 Stanley, Roy. *World War II Photo Intelligence*. pp. 116–127.

17 Babington-Smith, Constance. *Evidence in Camera*, pp. 50–51.

18 Wikipedia, Aerial Reconnaissance, https://en.wikipedia.org/wiki/Aerial_reconnaissance, viewed 24 February 2021.

19 C. Babington-Smith. *Evidence in Camera*, p. 177.

20 Map-film correlation is the painstaking process of comparing information on photographs to the most recent map available of the region that was photographed. In-scene scaling enables an interpreter to use a known dimension—the width or gauge of a railway track in a certain country to estimate sizes or distances on a particular image.

21 www.bbc.co.uk/history/worldwars/wwtwo/ff7_vweapons.shtml. Viewed 27 June 2021.

22 Stewart, Paul, Dr. *Medmenham: Anglo-American Photographic Intelligence in the Second World War, Volume 1* (Submitted for the Doctor of Philosophy at the University of Northampton, 2019), pp. 160–162. Stewart's analysis presents the most comprehensive account of the volume of work and the multiple priorities in June and July 1944.

23 Klein, Gary. *Seeing what Others Don't: The Surprising Ways Humans Gain Insight* (New York, NY: PublicAffairs, 2013), pp. 33–36, 42–43.

24 Prange. Gordon. with Donald M. Goldstein and Katherine V. Dillon. *At Dawn We Slept: The Untold Story of Pearl Harbor* (New York, NY: Penguin, 1982). While the Japanese did not have aerial photography of Pearl Harbor, they had routine ground observations from 27 March 1941 through 6 December 1941. Also, their primary source of intelligence in Hawaii, the Naval intelligence officer, Takeo Yoshikawa, did take an aerial tour of the islands in the fall of 1941 (pp. 254–255) to confirm the accuracy of his ground observations.

25 Perret, Geoffrey. *There's a War to Be Won: The United States Army in World War II* (New York: Random House, 1991), pp. 38–39.

26 Stewart, Paul. *Medmenham: Anglo-American Photographic Intelligence in the Second World War,* Vol.1, p. 108.

27 https://collections.leventhalmap.org/search/commonwealth:ht2507 838. United States. Office of Strategic Services. Research and Analysis Branch. "Germany, Concentration Camps (June–1944)." Map. Washington, D.C.: OSS, 1944. Norman B. Leventhal Map & Education Center (accessed 27 May 2022).

28 Brugioni, Dino A. and Robert G. Poirier. "The Holocaust Revisited: A Retrospective Analysis of the Auschwitz-Birkenau Extermination Complex," *Studies in Intelligence*, Vol. 22, No. 4, Winter 1978, pp. 11–31.

29 http://auschwitz.org/en/history/auschwitz-iii/ig-farben, viewed 6 December 2021. I.G. Farben located the facility at Auschwitz to avoid Allied bomber attacks and to take advantage of the forced labor available from the nearby camps.

30 Brugioni and Poirier, p. 21.

31 Brugioni, Dino. A. *Eyes in the Sky*, pp. 139–144.

32 The earliest reconnaissance satellites were film-based, and they returned physical copies of negative film from space. Starting in 1977, the KH-11 near-real-time satellite returned digital signals that were electronically processed and written onto photographic film that analysts could interpret. In the early 1980s, the technology developed so that a few analysts could look at the images on a first-generation electronic light table, the IDEX IA. The output from the film-return systems is called photographic, and from the digital systems, imagery.

33 Raymond L. Garthoff (1991) "SOVIET SNAFU: Case of the Wandering Radar," *Bulletin of the Atomic Scientists*, Vol. 47, No. 6, pp. 7–9. DOI: 10.1080/00963402.1991.114599909

34 Hoffman, *The Dead Hand,* p.321. The Soviet Union admitted that the radar was a treaty violation in 1989.

BIBLIOGRAPHY

http://auschwitz.org/en/historyikipediaz-iii/ig-farben.

Babington-Smith, Constance. *Evidence in Camera: The Story of Photographic Intelligence in the Second World War* (Phoenix Mill, UK: Sutton Publishing, 2004).

www.bbc.co.uk/history/worldwars/wwtwo/ff7_vweapons.shtml.

Brugioni, Dino A. *Eyes in The Sky: Eisenhower, the CIA, and Cold War Aerial Espionage* (Annapolis, MD: Naval Institute Press, 2010).

Brugioni, Dino A. and Robert G. Poirier, "The Holocaust Revisited: A Retrospective Analysis of the Auschwitz-Birkenau Extermination Complex," *Studies in Intelligence*, Vol. 22, No. 4, Winter 1978, pp.13–31.

Burtch, R. Class Notes Sure 340, 2009, *History of Photogrammetry*, Ferris State University. pp. 21–22, viewed 19 March 2023.

collections.leventhalmap.org/search/commonwealth:ht2507838. United States. Office of Strategic Services. Research and Analysis Branch. "Germany, Concentration Camps (June–1944)." Map. Washington, D.C.: OSS, 1944. Norman B. Leventhal Map & Education Center.

Conyers, Roy Nesbit. *Eyes of the RAF: A History of Photo-Reconnaissance* (Washington, D.C.: Sutton, 1996), pp. 54–58.

https://creativecommons.org/publicdomain/mark/1.0.

Finnegan, Col. Terrence, J. *Shooting the Front: Allied Aerial Reconnaissance and Photographic Interpretation in the Western Front – World War I* (Washington, D.C.: NDIC Press, 2006).

Gallison, Peter. *Einstein's Clocks, Poincare's Maps: Empires in Time* (New York, NY: Norton, 2003).

Garthoff, Raymond L. "SOVIET SNAFU: Case of the Wandering Radar," *Bulletin of the Atomic Scientists*, Vol. 47, No. 6, pp. 7–9. DOI: 10.1080/00963402.1991.114599909.

www.gislounge.com/john-snows-cholera-map-gis-data/.

Goddard, George W. *Overview: A Life-Long Adventure in Aerial Photography* (Garden City, NJ: Doubleday, 1969).

www.theguardian.com/news/datablog/2013mar/15/john-snow-chol era-map.

Hoffman, David E. *The Dead Hand: The Untold Story of the Cold War Arms Race and its Dangerous Legacy* (New York, NY: Doubleday, 2009).

Johnson, Steven. *The Ghost Map: The Story of London's Most Terrifying Epidemic, and How It Changed Science, Cities, and the Modern World* (New York, NY: Penguin, 2008).

Klein, Gary. *Seeing What Others Don't: The Remarkable Ways We Gain Insights.* (New York, NY: PublicAffairs, 2013).

Lloyd, Steve. "Cotton's Covert Cameras: New Perspectives via Reverse Engineering," *Medmenham Magazine* , Spring 2020, pp. 22–25.

www.metmuseum.org/art/collection/search/283189.

Nadar, Felix, Gaspard Felix Tournachon and Thomas Repensek. "My Life as a Photographer," *October*, 1978, No. 5, Photography (Summer 1978), pp. 6–28.

O'Connor, Jack. *NPIC: Seeing the Secrets, Growing the Leaders* (Alexandria, VA: Acumensa Press, 2015).

www.ordnancesurvey.co.uk/about/history.

Perret, Geoffrey. *There's a War to Be Won: The United States Army in World War II* (New York, NY: Random House, 1991).

Prange, Gordon, W. with Donald M Goldstein and Katherine V. Dillon. *At Dawn They Slept: The Untold Story of Pearl Harbor* (New York, NY: Penguin, 1982).

Stanley, Col. Roy M. *World War II Photo Intelligence* (New York, NY: Scribners, 1981).

Stanley, Col. Roy M. *V Weapons Hunt: Defeating German Secret Weapons* (Barnsley, UK: Pen and Sword Military, 2010).

Stewart, Paul. "Medmanham: Anglo-American Photographic Intelligence in the Second World War, Volume 1." (Submitted for the Doctor of Philosophy at the University of Northampton, 2019).

Wilford, John Noble. *The Mapmakers*, revised edition (New York, NY: Vintage, 2001). www.arcgis.com/apps/Cascade/index.html?appid=392cd1e3f9364cf59405dd5ad3121bcd.

https://em.wikipedia.org/wiki/AerialReconnaissance.

CHAPTER 4

Recording

INTRODUCTION

From Da Vinci's time what had been envisioned, discovered, or comprehended about the world had to be recorded if it was judged to matter. Behind the precision of Di Vinci's bird's-eye view of Imola, a list of recorded measurements from his hodometer must have existed. In World War I, when aerial photography was first used as a source of intelligence, military intelligence analysts devised recording systems to capture the essential information. Initially, essential information dealt with where and what was photographed. Quickly essential information expanded to encompass how the information had been obtained. This information about acquiring the aerial photography, now called metadata, began the age of modern recording, the third intellectual activity of geospatial intelligence.

Throughout the history of each discipline that contributed to geospatial intelligence—cartography, photointerpretation, mensuration or measurement, and imagery analysis—recording followed observation. Until World War II, almost all recording comprised catalogues of photographs, the maps they were converted into, and words and numbers derived from observations.

During that war, recording changed in two ways. A British Air Intelligence officer, Douglas Kendall, created a system of sequencing the recording after developed film became available following an aircraft mission based on priorities that directed the analysts' attention. And the strategies developed by Allied commanders to defeat Nazi Germany led to a system of recording based on a potential target's value in light of those strategies. These two drivers for recording—immediate priority and future value based on strategic significance—would shape recording throughout the Cold War until digital imaging and visual recording began in the 1980s.

DOI: 10.4324/9781003436836-5

The volume of photographic intelligence information needed for future targeting drove one of the earliest US military computer acquisitions in the early 1950s. The subsequent reliance on computer technology for intelligence production has never ceased. The introduction of commercial geographic computer software and hardware, now called geographic information systems (GIS), in the 1980s, began to change the nature of recording. The use of computer graphics for capturing and visualizing information, computer storage for retaining information, digital networks for transmitting information globally, and computational speed for revising earlier information enabled and enlarged the possibilities of geospatial intelligence.

These technological innovations brought the capability to reuse recorded spatial information routinely, to locate quickly that which couldn't yet be characterized, and share geospatially accurate information around the globe rapidly. These new capabilities enlarged imagery analysis and cartography, and in the early years of the 21st century, enabled their transition into what now is called geospatial intelligence.

This chapter also addresses the computer-generated recording and the data problem caused by the persistent video surveillance enabled by the drone era in the 21st century.

RECORDING—THE WORK BEFORE THE REAL WORK

One little researched challenge for photointerpretation, imagery analysis, and geospatial analysis is the discipline of recording what has been discovered. After the thrill of discovery, geospatial analysts have the work of recording. Since Leonardo da Vinci's 1503 aerial map of Imola, only a percentage of man's recorded discoveries about the planet remain, but no unrecorded discovery remains. Even before Da Vinci created the overhead view, he or one of his assistants must have listed all the measurements he had made with his hodometer along the streets and walls of Imola.

Since the first extensive use of aerial photography in World War I, the challenge for those looking at film or images of the earth has been deciding what information from the photograph to record in words or numbers and what information to ignore. Since the onset of World War I, the economics and risk involved in aerial and satellite photography caused the information on the image to be translated into words for comprehension, transmission, visual rendition, or retention in a storage file or data base. The partial or complete translation of images into another storage medium is the art of recording.

In World War I, the "translation" requirement affected all combatants. Only some senior military officers had the ability to interpret, understand, or employ this new intelligence source. This ability was distributed unevenly and, in some cases, inadequately. In every World War I military service, senior officers had been trained from the beginning of their military careers to use paper maps to envision current changes in a battle or front and to plan future changes. Many military officers, in multiple countries, were skeptical initially about the utility of this new photographic source of information. This skepticism led to the routine daily practice across the entire front of transferring information and intelligence from aerial photography to a strategic map, what the French army termed the "*plan directeur.*"[1]

More than 400 years ago Shakespeare captured the challenge for intelligence officers when Othello, a military leader out of his element, demanded "the ocular proof,"[2] only not to comprehend fully what was presented to him. The "translation" requirement from photography to two-dimensional maps in World War I routinely delayed the production of intelligence for all the combatant nations, and it would continue to affect photointerpretation, imagery analysis, and geospatial intelligence into the 21st century. The real-life Othellos in World Wars I and II, and the Cold War, would repeat the requirement for "the ocular proof," from aerial and satellite photography. In every case the recording of information shaped the definition of proof and the utility of photo-, imagery-, or geospatial intelligence.

But recording costs time. From 1914 until the late 1980s, when GIS technology was first being introduced to the intelligence community, the communications challenges forced decisions about what information to record in another milieu—text, maps, or tables—and what information to leave on the film or image or in the memories of the analysts. Throughout this time, the rationale for every recording decision was verbal—what necessary information had to be translated from the visual medium of film to the verbal medium of text for discussions and decisions. Military leaders and civilian policymakers were accustomed to and experienced in receiving information in written text or maps. During this time civilians and military officers at all ranks increased their awareness of overhead photography and imagery, yet a number of them still had some difficulty envisioning and interpreting what was being shown to them.

THE BIRTH OF ANALOG METADATA

In geospatial recording, the same basic information is almost universally recorded in text. For an image to have utility, it has to be tied to a place, a time, a technology, a measure of utility, and a sequence. Today, these measures are commonly called metadata, or data about the geospatial data.

During World War I, the basic metadata were established and their utility continues today. The first category of metadata was temporal and locational. The time of the photograph mattered as the captured content in the photo would continue to change after the photograph was taken. Any information or intelligence captured in that slice of time was perishable, and perishability drove many process improvements during that war. The information about the location being photographed often was far less precise than the time. The earliest pilots and observers had only imprecise instruments on their aircraft to indicate their exact location, azimuth, airspeed, altitude, and angle relative to the terrain.

The technologies used to take the photographs and images also determined the accuracy of the attempted collection. This is the second category of metadata, and it incorporates information about the location of the aircraft. The early photointerpreters, in World Wars I and II, through the mathematics of photogrammetry, could combine the focal length of the camera, the linear dimensions of the print, and the altitude of the aircraft to measure the area of earth covered by the image or specific objects shown on the image. This mathematical process is called mensuration. The sequence of photographs and the flight azimuth could then provide an accurate track or coverage record for the mission.

Camera technology determined the utility of the photography and defined the third category of the metadata. Some cameras were designed to capture light over a wide field of view, and others were designed to photograph a smaller field of view at a higher magnification. The third common design was a camera that incorporated higher resolution than the area collector and a greater field of view than the high-resolution camera. Consequently, information about the camera—later in the digital era termed the "sensor," and the aerial vehicle—balloon, aircraft, drone, or satellite—commonly called the "platform"—became as necessary as the information about the earth.

The fourth category of metadata concerned transient conditions that could influence or preclude success. Depending on the latitude and time of year, the daily duration of sunlight limited the time available for aerial overflight and photography. Also, transient conditions—haze, fog, dust, smoke, or clouds—affected the photographs, and would have been recorded in metadata.

Finally, in the fifth category of metadata, the photointerpreters created measures of intelligence utility. This covered the human elements in the process—inaccurate pointing, partial coverage, target-off-frame, poor contrast, excessive or inadequate obliquity, camera out of focus, or insufficient resolution to answer the intelligence question.

Before an intelligence question could be answered, the metadata had to be captured. Only after recording could the intelligence value of the image or mission be determined. And as each intelligence report generated one or more requests for a follow-up, or additional reports, all aerial intelligence organizations created ways to file each photograph for re-use.

These recording decisions about converting and communicating the results of photo-intelligence, imagery analysis, or geospatial intelligence into prints, paper, or a database have been captured in every narrative of organizations responsible for analysis from these sources for a hundred years.[3] These operational requirements caused photointerpretation, imagery analysis, cartographic, and geospatial intelligence organizations to be resource intensive for both people and technology. Attempts to simplify and accelerate these processes shaped the operations of every one of these units.

MEASURING URGENCY

In July 1940, the British Royal Air Force for the first time established a model for dealing with the urgency of current intelligence and recording the essential information, also called time-dominant intelligence.[4] Douglas Kendall is given credit for establishing the three-phase model of exploitation and analysis to derive intelligence from aerial photographs. The landing time of the reconnaissance aircraft drove Kendall's model. Every mission was flown in response to priority questions requiring an answer or response within two hours after landing. Answering these questions constituted the first phase of exploiting the information on the photographs. Generally, these questions focused on operational planning, indications and warning of immanent attack, recent changes in the status of opposing forces, or the highest current interest issue.

Second-phase exploitation involved answering questions within the first 24-hours after landing. Most often, these questions related to current threat assessments, battle damage assessments from prior strikes or engagements, or strategically important issues such as the development status of new weapons systems. Frequently, answering these questions required additional review of prior photographic coverage of the target or area to provide context and determine the extent of change. The amount of available recorded information often determined the depth of the reporting.

Third-phase exploitation reviewed newly collected film or imagery for important questions that did not need responses within 24 hours. These questions required extensive re-searching of prior film or imagery collection. Often third-phase questions could be as important as time-dominant questions, as in the case of strategic weapons development and testing, but third-phase questions require researching and studying prior photographs to discern subtle changes. Kendall's model was important because first-phase, time-dominant intelligence often was communicated as quickly as possible through telephone, radio transmission, or text-only reporting. The recording was done after the initial reporting. Yet, Kendall knew the value of research and the potential of photo-reconnaissance to provide more than immediate answers. So, the first-phase images were recorded retrospectively. The subsequent two phases of the model allowed time for recording and research. The amount of recording defined the quality of the research. Photographic or imagery research is impossible without systemic recording of collection and prior analysis. Imagery or geospatial research can provide intelligence that no other analysis can achieve.

LOOKING BACKWARDS TO GET AHEAD

When analysts have time to review and research older images, frequently they can develop indications of future activity or signatures that allow analysts to confirm that certain types of equipment are at certain locations. The hunt for the V-1 and V-2 launch sites would be in this category as well as strategic weapons systems or processes, such as submarine construction, aircraft R&D facilities, or petroleum refining facilities. Kendall's model systemized Jean-Louis Weiller's World War I insights about the value of relooking at older photographs and keeping installation histories. The analytic successes of the British and American photointerpreters at ACIS in learning about the V-weapons, the German jet-propelled aircraft, and developments in submarine production and radar deployments could not have been achieved without this model.

Kendall's three-phase model endured long past World War II. When Arthur Lundahl set up the HTAUTOMAT organization inside CIA in 1956 to exploit future U-2 missions over the USSR, he used this model. When his organization became the National Photographic Interpretation Center in 1961, this model continued in use, and in 1977, with the onset of daily digital electro-optical satellite imagery in the Kennan, KH-11 program,[5] this model for recording was used again and continues in use today.

Along with Kendall's three-phase model, the Allies introduced two changes in recording geographic and spatial information during World War II. One type of physical recording was created for a single event and the other for a longer duration campaign. As part of the preparations and training for the Normandy invasion, a team of model makers augmented the photointerpretation group at Medmenham.[6] Modelmaking for individual targets using aerial photography as a source had been ongoing since the beginning of the war, but the creation of multiple copies of more than 100 terrain models over the area of the seaborne and airborne invasion was a new endeavor for the model makers. These artists and craftsmen created hundreds of copies of these terrain models, based on aerial photography, showing critical objectives for each unit in the invasion.[7] The invasion forces used these models before the landings to familiarize themselves with the locations, their surrounding terrain, proximity to other landmarks, and unit objectives. The training based on these models provided the familiarity that enabled, in part, the first Allied invasion units to orient themselves more rapidly than the German defending forces. The use of models to explain imagery and geospatial analysis and to assist in operational military planning continues into the 21st century.

The other recording innovation, the creation of a formatted future target encyclopedia, was part of the Allied strategic bombing campaign in Europe. Each of these innovations would help the future development of geospatial intelligence recording. The target encyclopedia for Allied bombing marked the rudimentary beginning of the first formatted data base for geospatial information. As part of an effort to determine the importance of selected German military infrastructure, to rank potential bombing targets, and to diminish the risks to Allied airmen on daytime and night missions, the Allied Air Forces created the Bombing Encyclopedia. The Bombing Encyclopedia provided a way for the Allied Air planning staffs to record data about known and potential bombing targets and to record the essential elements of information about any potential targets.[8]

Most of the primary information in the Bombing Encyclopedia was geospatial—the location, size, boundary, and orientation of the target, as well as its relation to nearby landmarks. Much of the additional information was military, including the most valuable aim points at a target, known locations of the near and far air defenses, the orientations and possible chokepoints that might be blocked along lines of communications to the target, as well as nearby secondary targets. And the final category of information covered any nearby areas that were not to be bombed, such as schools, hospitals, or culturally significant sites. In the European theater in World War II, the Bombing Encyclopedia served as a mission planning tool for the Allied Air Forces, but its postwar value turned out to be greater.

THE ENCYCLOPEDIA MEETS THE COMPUTER

In the early years of the Cold War, the World War II bombing encyclopedia was modified and connected to a new technology. During the 1950s, as the limitations of long-range bombers at penetrating either the Iron

Curtain around the USSR or the Bamboo Curtain around China became evident, the bombing encyclopedia was renamed the Basic Encyclopedia (BE). This became the first digital database for strategic Cold War targets. More importantly, the Basic Encyclopedia provided the first format for recording metadata digitally about individual targets of interest. And, as in World War II, most of the information and metadata recorded in the Basic Encyclopedia was derived from photographic intelligence.

The BE changed the fundamental nature of geospatial recording for the first time since 1502. Before its creation, the limit on the information about how aerial photographs were obtained had been human memory and human computational accuracy. This limit changed in 1952 when the second Univac computer purchased by the US government went to the Air Force to maintain the BE.[9] The computer memory enhanced the BE's accuracy and accessibility. While the quality of its records would remain dependent on human limitations for new information about the earth, the computer removed the limits of human memory from managing the recording process.

Before the creation of the BE, the information management or recording question had been: what can humans record? With the creation of the BE, the question became: what data could computers record and retain to diminish reporting responsibility for the humans who looked at photography? The BE could store only formatted text. Each individual target of interest received a unique BE number. Targets began as a hand-lettered entry keyed into a computer through punch cards. The recently created US Air Force managed this initial digital toehold on the meta-data problem. The information related to the BE number remained text-only so the verbal and visual information about these targets of interest remained separate, as there was no way other than text to introduce map or photographic information into the BE.

By 1958, after two Soviet Sputnik launches from Tyuratam, the computer data base had become an integral part of US strategic thinking about intelligence and military targets. At that time the US began to make a very large investment, which continues today, in photo-satellites, imagery analysts, and military intelligence organizations to locate, identify, and record Soviet, now Russian, and Chinese military capabilities and infrastructure.[10]

From 1954 through 1974, incremental gains in computer processing speed and computer memory improved the recording of geographic and military information from photo-satellites. Yet while the analysts' textual reports were being stored in computers and communicated electronically, their photographs, coming back from space in increasing numbers, remained trapped by their silver halide emulsion in countless film cans shelved throughout the Intelligence Community and the Defense Department.

But the recording challenge remained a function of analytic attention. When organizations have focused the information and attention of their analysts, as Western intelligence and cartographic organizations did during the Cold War, they had deep records about human changes on those parts of the planet. But since the Cold War the growing volume of government and commercial imagery and the proliferating number of small imagery satellites exceeds the available human attention. Today, the number of imaging satellites on orbit can create more imagery than millions of humans can look at. Yet, in combination with neural-network algorithms, this new space technology can assist in a rapid human discovery of knowledge and awareness of areas of the planet not previously attended to.

VISUAL RECORDING

The inability to use computers to store geographic data in any form but written text constrained cartographers and photointerpreters until the early success of the American LANDSAT remote sensing digital satellite in July 1972.[11] LANDSAT proved the concept of digital imagery transmitted from space, and created the issue of how to store large files of digital satellite data. Less than five years later, the first classified US digital intelligence satellite—the Kennen KH-11—was launched in December 1976 and began to operate in January 1977.[12] The initial years of digital satellite operations presented a time of opportunities and challenges. For the first time reliable observations in the unclassified and classified modes could be produced, but most of the collected digital images had to be converted to hard copy film to be analyzed.

By the 1970s, the earliest work on commercial digital GIS had begun at Intergraph and Environmental Systems Research Institute (ESRI) in the US and at Siemens in Germany. Conceptual work on GIS had started in the 1960s with the experimental Cartography Unit in the United Kingdom, the Canada Geographic Information System for the Agricultural Research Development Agency, the Harvard Laboratory for Computer Graphics, and the Swedish Land Information System (LIS).[13] GIS technology would eventually change the nature of graphic recording and play a critical role in the creation of geospatial intelligence in the 21st century. But GIS did not begin to penetrate the US Intelligence Community until the 1980s. Initially CIA used GIS to prepare accurate thematic maps and unclassified line drawings based on classified imagery of Soviet installations. US State Department arms control inspectors used these line drawings to make onsite inspections in the USSR in support of the INF and SALT II disarmament treaties.[14] These GIS-based line drawings enabled the US to protect the classified satellite imagery sources from which the drawings were made. Through the 1960s and 1970s, the increased volume of film and images from reconnaissance satellites drove the recording technologies for imagery analysis and cartography. Only in the 1990s did digital imagery and online storage begin to change how most imagery analysts worked. Digital imagery enabled new kinds of recording and the incorporation of GIS technology greatly accelerated intelligence production and communication. GIS connected digitally the targets that had been filed individually in text reports and paper maps. It allowed analysts to keep and store the relations among targets spatially, digitally, and precisely.

In the US Defense Mapping Agency (DMA), responsible for Defense Department cartography since 1972, the path to digital technology was as arduous and slow as in the imagery intelligence community. During the 1980s, their initial proprietary GIS system for digital mapping, the Mark 85, had been plagued by delays and cost overruns. Its replacement, the Mark 90, also delayed, came online in 1992,[15] and DMA's initial experimental use of commercial nonproprietary GIS software began in the late 1990s to early 2000s.

In the US Intelligence Community, the divide between recording the visual and recording the verbal slowed improvements in imagery intelligence analysis and communications. Throughout the 1980s and early 1990s, while digital imagery analysis was full of promise, the available exploitation technology lagged behind. In the 1970s and 1980s workstations to process and display digital imagery had to be custombuilt at a cost of several hundred thousand dollars each.[16] The computer memory necessary to store even just a few images was then prohibitively expensive. The digital imagery satellites accelerated the pace of the development of these technologies, but it would be in the middle of the 1990s before the digital combining of words and pictures from digital satellite imagery happened routinely at the workstations of most imagery intelligence analysts. It would take an existing US government space technology, GPS, as well as the commercial development of the personal computer, along with the growth of commercial GIS software to provide the remaining technologies that enabled the transformation of digital imagery analysis and digital cartography into geospatial intelligence.[17]

These final two essential technologies arose from opposite requirements. During the Cold War, the US Department of Defense needed to solve the problem of targeting nuclear weapons accurately from land and sea. The mobile nature of submarines and the mobile basing plans for ground-based missiles required that the missile launchers define their precise locations before being able to program the weapons to strike targets accurately. The GPS solved this targeting problem and the GPS capability to define precise locations also created other commercial possibilities.

The earliest satellites in the global positioning system went on orbit in 1978. President Reagan authorized nonmilitary use of GPS in September

1983 when the Soviet Union shot down Korean Air Lines flight 007. With media reporting about the US military use of precision-guided weapons in Operation Desert Shield/Desert Storm in 1991, greater public awareness of GPS came about. After this first public display of the military potential of GPS, President Clinton in 2000 released the restrictions on GPS technology and the civilian marketplace began to create applications for this technology.[18]

While computer technology had been helping humans record more accurately what was changing on earth, the effects of GPS were not integrated into the film-return satellites in space in the 1980s. The film-return systems in the 1980s KH-8/GAMBIT 3 and KH-9/HEXAGON, like the earlier systems (KH-7/GAMBIT 1 and KH-1-4/CORONA), all relied on star tracker technology for positional accuracy. While this was the best available technology at the time, hardware or software issues in space would frequently degrade the pointing accuracy of the film-return satellites. And when the hardware failed, imagery analysts would frequently have to report—"target off frame," or "partial coverage indicates." And the cartographers could only throw up their hands.

From the onset commercial digital imaging satellites adopted GPS to improve their accuracy, and these accuracy issues disappeared. In the very late 20th century and the early years of the 21st century, GPS ended the reliance on humans for metadata recording. The computer, due to GPS, provided reliable and accurate locational metadata. By this time, the growth and improvement in government satellites, and the growing number of commercial imaging satellites, caused the volume of geospatial images to increase and strain the analysts' capabilities to look at all the imagery.

During the 1980s and 1990s GIS achieved maturity and began to be used in many sectors of society. The genesis for GIS was not military. GIS made digital recording of the visual aspects of cartography and imagery analysis as easy and repeatable as the verbal aspects. This technology moved from proprietary hardware systems in the 1980s to software that could be hosted on multiple kinds of workstations and personal computers in the 1990s. A private company in Redlands, California, ESRI, drove this change, and it began to dominate the commercial market.[19]

GIS enabled geographers and intelligence analysts to record changes accurately in three dimensions, and to revise these changes at the speed of satellite sampling and digital computation. The ability to use software to perform computational geography with digital accuracy solved the visual recording and revising challenge that had endured since Da Vinci's map of Imola. Not until the 21st century would the full effects of GIS be felt globally when commercially available digital imagery and Google Earth made locational information available for anyone with an Internet connection.[20] The civil and unclassified geographic communities had discovered the utility of GIS well ahead of the intelligence community. The ability of anyone with Internet connectivity and a sufficiently powerful computer to command the technologies of digital commercial satellites, GIS, GPS, and individually customizable imagery analysis and cartography enabled and democratized the creation of geospatial intelligence. The rapid expansion of these intelligence and defense technologies changed exclusively governmental work into individually accessible geospatial intelligence. GIS technology brought digital accuracy to visual recording. Accurate locational information and computer-assisted analysis of recorded data have become routine.

SPEED, DIPLOMACY, AND ATROCITY

Over the past 25 years, the rapid growth in GIS, computer memory, and processing speed enabled and accelerated the growth of geospatial intelligence. The first notable use of GIS to accelerate intelligence recording and production was its incorporation into disaster responses in the early 1990s. This was done to create accurate damage assessment maps rapidly, and to avoid having to share classified satellite images with audiences not usually cleared to view them. The speed of the response was made possible by GIS. Later in the 1990s, military and diplomatic requirements for intelligence about the wars in the former Yugoslavia would begin to push imagery analysis and cartography closer.

The GIS and satellite imagery capability to track day-by-day the ethnic cleansing and the actions of Croatian, Bosnian, and Serbian forces enabled the display of the pattern of selective destruction of Muslim communities and exposed the Bosnian ethnic cleansing. The advantages to recording change in a GIS are now considered commonplace—rapid revision, precision location, global transmission over the Internet in shape files with new or existing layers of information, and the ability to perform statistical analyses of the patterns of change to enable more detailed and more accurately predictive spatial analyses. Like GIS disaster support products, the daily fire maps of Kosovo were releasable to all parties without disclosing imagery satellite capabilities. The use of GIS-based maps for diplomatic release to the involved parties, as well as the public, the media, and eventually the World Court, changed intelligence, diplomatic, and public awareness of the events and the war crimes among them. The accuracy of the maps, presented to all parties in the negotiations, convinced the parties of the reliability and the objectivity of the information. On account of the Cold War, the US had recorded considerable knowledge in its databases about the military and the terrain in the former Yugoslavia that it incorporated in this analysis. Within a decade, in a far different and more desolate place, the use of GIS for recording imagery analysis exposed another atrocious war crime against humanity.

In 2004, a National Geospatial-Intelligence Agency (NGA) imagery analyst returning from military reserve duty was assigned to work on the Sudanese military, an issue unfamiliar to him. In his discovery of the status of those forces, this analyst noticed Sudanese military equipment in a region of western Sudan called Darfur. Working from unfamiliarity on a previously inactive account with little recorded information, the analyst could only record his initial observations by latitude and longitude. But his observations of battle damage far from the expected area of fighting in the Sudanese Civil War made him curious. His initial investigation took two parts. The analyst collected current imagery over large areas of western Sudan, and discovered many more damaged or destroyed villages. As no US maps existed with place names for this region, the imagery analyst began to create a spreadsheet to record his observations from which a geospatial analyst could begin to build an accurate GIS to capture the extent of the damage.

After significant effort, as Darfur is as large as Texas, the analyst and his small team established a pattern of ethnic cleansing by the Sudanese government against the people of the Dar. The compilation of the atrocities over time became compelling. From the GIS which was used to record all the damage reporting, NGA created a map that became the basis for Intelligence Community briefings to Congress and to the 2004 US Department of State declaration of the events in Darfur as genocide.[21]

PARTICIPATORY AND VOLUNTARY RECORDING

In the 21st century, the availability of commercial imaging satellites and accessible geospatial technology led to the rapid creation of volunteer groups around the globe to capture and record geospatial information. The earliest of these voluntary groups formed after the Haitian earthquake on 12 January 2010.[22] After the earthquake struck, little geospatial information and less context were available to local and international search and rescue forces, as Haiti is a poor country that could not command enough attention for foreign countries to devote resources to mapping it accurately. Specifically, the search and rescue forces needed cadastral maps of Port-au-Prince. Cadastral maps show street names, address numbers, and sometimes the names of property owners.

Multiple nonprofit organizations—Open Street Map, Ushahidi, CrisisCampHaiti, and GeoCommons,[23] reached out globally to anyone with information about Haiti, specifically the Port-au-Prince area. In many countries, Haitian expatriates and others responded. Through the digital technology of the Internet, GIS, and cell phones, these groups used a GIS to spatially organize the data and created a series of maps quickly enough to assist in search and rescue missions as well as later volunteer recovery missions.

A similar use of volunteer recording occurred in 2011. Skytruth, a small geospatial intelligence organization in Shepherdstown, West Virginia, wanted to locate all the fracking locations in Pennsylvania[24] and Ohio.[25] Fracking involves drilling into the earth and injecting water and chemicals under pressure to release natural gas. The customer for its search was the Bloomberg School for Public Health at Johns Hopkins University. It wanted to plot fracking sites in the eventuality that future ground water contamination became a regional public health issue. Uncertainties remain about the effects on the water table of the fracking chemicals pumped underground at high pressure. Skytruth, a small organization, with fewer than 10 employees, understood geospatial analysis very well, and created a plan to solicit and train volunteer imagery analysts to search satellite imagery of Pennsylvania and Ohio on Google Earth. Those volunteers who passed the training were given imagery coverage over an area to look for fracking sites or potential fracking sites.

The results of the initial volunteer search—annotated commercial images geospatially registered to the locations where amateur imagery analysts observed ground scarring and other activities–enabled the small staff at Skytruth to identify the actual fracking locations. This analytic effort was accomplished quickly because Google Earth and GIS had already provided a considerable part of the necessary metadata. The search results, recorded in an ESRI ArcPro database, remain visible to the general public.[26] Since that time, Skytruth has been incorporating machine learning into identifying fracking sites.[27]

Crowd-provided metadata helped identify the Russian military criminals who shot down a commercial Malaysian aircraft, MH17, over Ukraine on 17 July 2014. Bellingcat,[28] an independent investigative organization, combined commercial imagery with geospatially registered cell phone videos and still photos, mostly provided by volunteer researchers. Through patient searching, tracking, comparing, analyzing, and recording imagery and geospatial information, the Bellingcat investigative team compiled and documented a narrative timeline of the shootdown. Their geospatial analysis identified the location and the specific Russian surface-to-air missile transporter-erector-launcher (TEL) that launched the missile that destroyed the aircraft.

ALGORITHMIC RECORDING: COMPUTERS
START TO LOOK

In the late 1950s and early 1960s, two intelligent contributors to the history of geospatial intelligence had been thinking how computers could assist photointerpreters and what computers might be able to identify on photographs or images. One needed relief for his overworked interpreters, and the other envisioned how software could teach itself to help photointerpreters by recognizing and recording what was shown on the image.

In the late 1950s, the potential of computers and the limits of human beings weighed on the thinking of Arthur Lundahl and Frank Rosenblatt. Lundahl, at CIA's Photo Interpretation Center in the Steuart Building, saw that the photointerpreters were already under the pressure of an accelerating assembly line of U-2 aerial photographs, much like Lucille Ball in the historic chocolate-factory television skit,[29] but with far more serious consequences. After the receipt of each mission, Lundahl's small team of photointerpreters coped with the volume of U-2 intelligence with energy and continuous overtime. Lundahl knew that his workforce was going to be increasingly challenged. As he was already cleared for the CORONA photographic satellite then under development. At some future date relying on overtime and the energy of the photointerpreters would not be a sustainable plan for producing timely intelligence.

As Arthur Lundahl was coping with the demand side of the information management problem, a psychologist at Cornell, Frank Rosenblatt, envisioned how computers might replicate what humans were capable of doing with their eyes and minds. Specifically, Rosenblatt envisioned how to teach computers to learn to make visual discriminations and how to improve their accuracy with repetition. Rosenblatt had gotten the Navy to fund his experiment in 1958. To test his first assisted target recognition software program, the Perceptron, he built a prototype computer and sensor configuration, the Mark I Perceptron (Figure 4.1).[30]

To demonstrate the utility of his idea, in 1958 and 1959, Rosenblatt focused on computer identification of man-made objects.

In October 1960, Lundahl took the first step in the long developmental journey of geospatial image feature-recognition algorithms. He reallocated some of his current budget to initiate research and development on this issue.[31] When Lundahl took this step, only one successful photo-satellite mission had flown two months earlier in August 1960. The next successful photo-satellite mission would not occur until December 1960. But Lundahl's four years' experience with U-2 missions over the USSR and China, and his awareness of the eventual volume of

FIGURE 4.1 The Mark One Perceptron–the first feature recognition algorithm.

Source: National Museum of American History & Smithsonian Institution Archives.

satellite information that would come from the CORONA photo-satellite, persuaded him to invest in this experimental activity.

By spring 1963, Lundahl and his organization, now called the National Photographic Interpretation Center (NPIC),[32] had achieved acclaim for their efforts at warning about and then monitoring the resolution of the Cuban Missile Crisis from October through December 1962. In January 1963, NPIC moved to a much larger building and had undertaken a hiring surge to help look at and analyze the influx of satellite images. Rosenblatt, four years after having proven his concept about the single-layer Perceptron making visual distinctions and improving its accuracy, had not found a practical naval use for his experiment, and the Naval Research Laboratory was cutting his funding.

Rosenblatt's experiment caught Lundahl's attention, who thought that it might help his photointerpreters. In April 1963, Lundahl and his technical advisor, John Cain, persuaded CIA to fund an additional Perceptron contract. At this time Lundahl's organization had received and interpreted only 18 KH-4 CORONA missions with fewer than 30 days' worth of imagery. However, during the previous fall, during the Cuban Missile Crisis, NPIC analyzed more than 100 low-altitude reconnaissance aircraft missions in fewer than 60 days.[33] Lundahl knew then the frailty of the overtime model as well as the importance of developing computer assistance for the interpreters.[34]

The NPIC experiments with the Perceptron in the next few years continued to show promise, but the Perceptron was a slow learner. Its hardware and software improved its ability to correctly detect objects such as ships and aircraft, but it could not operate rapidly enough to help NPIC photointerpreters who could not wait for the Perceptron to train itself. NPIC ended its support to the Perceptron contract in 1967, and, through the US Navy, the Mark I Perceptron was donated to the Smithsonian in 1968.

The initial Perceptron experiments were successful but impractical. Thirty-five years later, Lundahl's investment paid off. Rosenblatt's single-layer Perceptron had been successfully developed into a multi-layer Perceptron. After many years of academic neglect, and much improvement in computer calculating capacity, speed, storage, and software complexity, computer scientists started building on Rosenblatt's idea. Success began to arrive in 2003 with the work of Geoff Hinton and Yann LeCun in Toronto. Their development resulted in the AlexNet algorithm that won the ImageNet competition at Stanford in 2012.[35]

The narratives of those involved in developing neural network algorithms for identifying features from images all outline the difficulty of the process and the need for extensive image libraries to "train" the algorithms. Two examples illustrate the complexity of the challenge: one from an extensively wealthy corporation, Google, and one from Skytruth, a small organization with a handful of employees who made use of free tools provided by Google.

In 2016–2017, in support of a Stanford project to use visual geo-registered information to identify voting patterns, Google attempted to write an algorithm that would distinguish and identify all the cars and trucks that appear on all the images in Google Street View. The project examined 50 million images and their location data. According to Timnet Gebru, the research leader who commissioned the Stanford effort: "Collecting and labelling a large data set is the most painful thing you can do in our field (Artificial Intelligence). But without experiencing that data-wrangling work, you don't understand what is impeding progress in AI in the real world."[36]

Skytruth undertook a similar feature recognition effort, on a much smaller scale. This nonprofit organization had successfully trained hundreds of volunteers in 2011 to look for fracking drill well pads in Pennsylvania and Ohio. From 2018 to 2019 Skytruth used the results of that earlier analysis to create a machine-learning model. Their algorithmic model was based on 8,000 images of well pads and "not well pads" compiled from a data set of 10,000 National Agricultural Imagery Program images.[37] The Skytruth effort illustrates the challenges and effort involved in creating a recognition algorithm. After "training" the software with 160,000 examples from this data set (4,000 pads + 4,000 not pads × 20 cycles), the article summarizes their results: "Our best model run returned an accuracy of 84%, precision and recall measures of 87% and 81% respectively, and a false positive rate and a false negative rate of 0.116 and 0.193, respectively."

As Ry Covington, creator of this model, points out:

In nearly every training example, there is a clearly defined access road that connects to the well pad. As a result, the model regularly classified large patches of cleared land or isolated developments (e.g. warehouses) at the end of a linear feature as a well pad. Another major weakness is that our model is also overly sensitive to active well pads. Active well pads tend to be large, gravel squares with clearly defined edges. Although these well pads may be the biggest concern, there are many "reclaimed" and abandoned well pads that lack such clearly defined edges. Regrettably, our model is overfit to highly-visible active wells *(sic)* pads, and it performs poorly on lower-visibility drilling sites that have lost their square shape or that have been revegetated by grasses.[38]

For all the effort, continuing development, and improvement invested in feature recognition algorithms since 1957, serious considerations remain. The algorithms, wholly reliant on prior imagery for their "training," will not succeed in identifying any new or unprecedented object. Some algorithms, as the Skytruth article points out, score higher accuracy rates when the area under consideration has more clearly defined geometry. Unlike humans, even very good feature recognition algorithms have difficulty in dealing with incomplete patterns. Two images illustrate the difference that mathematics and coding cannot overcome.

In the late 1950s and early 1960s, two images, one created by an Italian psychologist and the other from the camera of a classified U-2 mission, were seen for the first time and separately discussed. An intellectual construct created in 1955 by Gaetano Kanizsa, an Italian psychologist, demonstrates the challenge for neural network algorithms. This illustration (Figure 4.2) of the Kanizsa paradox presents the interpretative challenge of "describing what is known about the stimulus rather than what is seen."[39]

Kanizsa's triangle illustrates the strengths and weaknesses of human and computer discovery. Simple for humans to see can be atrociously

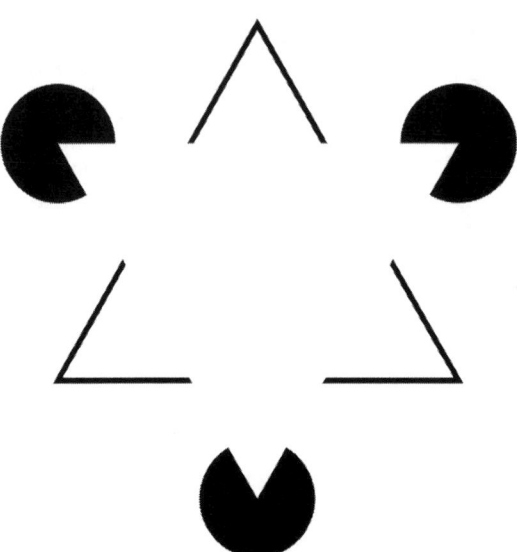

FIGURE 4.2   Kanitza's paradox.

Source: Fibonacci, Wikimedia (CC BY-SA 3.0).

difficult for the computer to detect. As Kanizsa demonstrated, the human brain can easily envision missing contours or textures.[40] Objects that Euclid could describe remain easier for computer algorithms to detect accurately. The more imperfect the geometry of a figure, the more difficult it becomes for the algorithms. When describing an object on an image requires non-Euclidean geometry, the algorithm's efficiency suffers.

Shortly after Kanizsa was studying how humans perceive, the Soviet Union, in response to the US U-2 overflights, built and deployed the high-altitude SA-2 surface-to-air missile system in 1957.[41] It was first observed by US photointerpreters in 1959.[42] The US photo-interpreters were able to identify the deployed SA-2 easily as its launchers were arrayed in a distinctive six-point star pattern around the radar and generating vehicles in the center of the star and cable connected to each launcher. The example, photographed in Cuba in 1962, illustrates the pattern (Figure 4.3).

From the perspective of photointerpreters, then and now, the SA-2 deployment pattern is easily recognizable and distinguishable from any other type of missile deployment.[43] The geometry of the Cuban SA-2 site is not perfect. While the circle is nearly precise, the triangular features are not, but for trained human analysts the geometric imperfections pose no difficulty in interpretation.

The distinction between human visual acuity and algorithmic detection hinges on precedent. For discovering, comprehending, or tracking

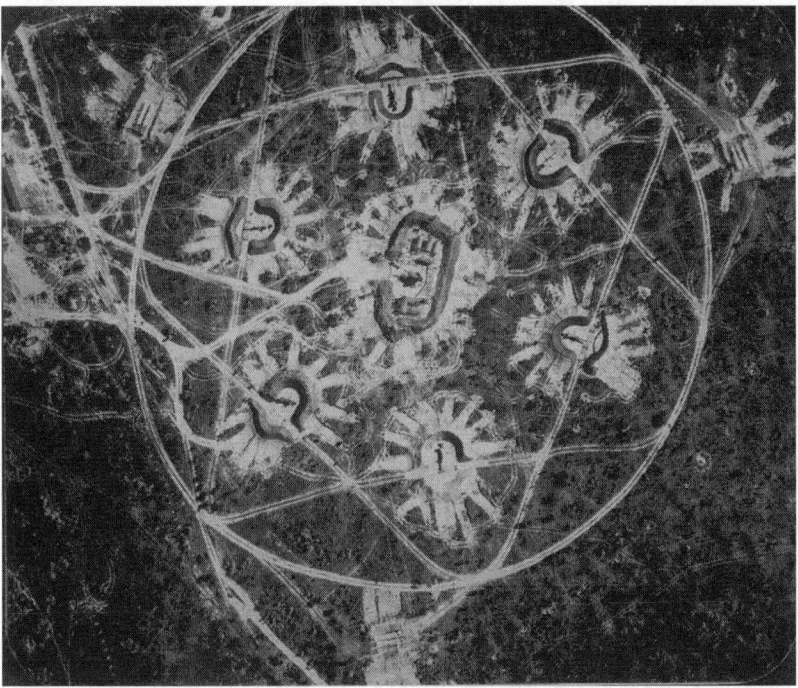

FIGURE 4.3   U-2 photograph of SA-2 Surface-to-Air Missile site in La Coloma, Cuba 1962.

Source: United States. Department of Defense. Department of Defense Cuban Missile Crisis Briefing Materials. John F. Kennedy Presidential Library and Museum, Boston.

activities or objects that are incongruous, partially missing, incomplete, or contradictory, human analysts remain better than algorithms at recording content. Perhaps the best example of this human capability involved the Soviet Union and Cuba.

During the Cold War, even before the Cuban Missile Crisis, the Soviet Union rotated military units in and out of Cuba. Later, during the Carter and Reagan administrations, the size and number of these units in Cuba became a contested US political issue. The intelligence question about the size of the reported "Soviet Brigade," which at the time could not be determined from available intelligence sources, provides an example where imagery analysis supplied essential information unobtainable from other sources.

Had neural network algorithms been available during the Cold War, it is very likely that the algorithms would have detected Soviet passenger ships at Cuban commercial port facilities. The geometry of the port facility and the geometry of merchant ships are sufficiently precise for algorithms to recognize distinctions among ship types and between the ships and the quay. Yet the judgment of an NPIC analyst, who worked on this issue during that time, illustrates a significant limitation of algorithmic reporting.

Tom Maddox, a DIA imagery analyst assigned to NPIC, worked in the Cuba branch. In his review of Cuban port facilities, he observed and reported on the transport activities of Soviet merchant and cruise ships. The Soviet Union had used cruise ships to bring Russian tourists to Cuba. As a deceptive measure, they used the same cruise ships for troop rotations between the USSR and Cuba. This practice made it difficult for intelligence analysts to assess the number of Soviet troops being rotated in and out of Cuba. After some imagery research, Maddox determined a way to identify when the cruise ships were being used for military purposes.

He made this distinction by noticing what was absent on the images rather than anything that was present. While the Soviet cruise ships had been used for troop rotations for some time, not until Maddox had relooked at all prior images of Soviet Cruise ships docked in Cuba that he noticed that on some images the swimming pools on the ships were filled with water and on others they were empty. Maddox built a chronology of when the cruise ships with empty swimming pools had arrived in Cuba and departed for the Soviet Union, and he was able to correlate these observations with other intelligence to confirm the timing and scale of the Soviet troop rotations.[44]

This analytic discovery illustrates a point made in imagery analyst training. For imagery and geospatial analysis, it can be as important or more important to notice what is absent from the imagery as it is to observe what is present. This point also bears on the development of algorithmic reporting through feature identification software. In the past decade, the 60-year effort had begun to bear fruit in identifying and recording what features are present on an image. With the growth of computer storage in the cloud, advances in processing speed, and the vast growth in commercial digital satellite imagery, feature recognition algorithms now have sufficient data to train themselves quickly enough to recognize objects with sufficient statistical validity to become useful to geospatial analysts.[45]

Due to advances in machine learning, algorithms have developed from recording information about the image to recording information from the image. Among the algorithms developed in the past decade are those that, with a high degree of accuracy, can identify, distinguish, and count military vehicles. Other algorithms distinguish the heights of floating-roof oil-storage tanks, and by measuring the variations on a daily basis, compute the fluctuations in the global petroleum storage market.[46] Other algorithms count and measure fixed structures, such as houses, over large areas. These algorithms have been used in disaster support efforts and in tax assessments.[47] While the algorithms, with sufficient commercial satellite and smallsat data, eventually will be able to record and make verifiable judgments about more complex terrestrial geometries, they may never be able to detect, as Tom Maddox did, when something absent from the image may be more important than all the objects that are on the image.

THE CHALLENGE OF MOTION IMAGERY

After 9/11 the US Defense Department has collected an immense amount of video drone data over Afghanistan and Iraq. In 2009 alone, as part of intelligence gathering against ISIS and the Taliban,[48] US Air Force drones collected more than 24 years of video recordings in Iraq and Afghanistan.[49] To accelerate the recording and exploitation of all this data, the US military developed Project Maven and began to work with Google to address this recording challenge, but the effort of recording all this video data highlighted two issues: operational and ethical.

Training algorithms to capture video data remained an operational challenge. In operational use, the algorithms had to be "trained" continually to diminish the rate of false positives and false negatives. Data access in combat differs from data access in orderly societies. Eric Schmidt, one of the leaders at Google, also chaired the US Defense Innovation Advisory Board, and described the contrast:

> these algorithms, at least today, require a great deal of training data. And when I say a great deal I mean like millions of entities in the matrices, billions of pieces of data. So the classic example: people would say "well why can't you figure out terrorism?" Well the good news is terrorism is very rare. Right? So it's much, much harder, if you will, you apply AI (artificial intelligence) to that problem. Whereas trying to understand traffic, right? As an example something that occurs every day, is far, far easier because you have so much training data.[50]

But recording by military analysts was also limited as these analysts were in Iraq and Afghanistan on rotational assignments, and their gained expertise would leave with them at the end of their tours. The military addressed the fluctuating knowledge issue by hiring civilian contract expertise, but the civilian contractors did not have equal access to information as the military.[51]

The ethical issue, using software to record and track human targets, who might be killed at the end of their tracking, emerged in 2019 and caused Google to cease work on Project Maven as some of its employees were resigning in protest. The Google employees were unhappy with being involved in automated targeting by the US military. When Google ceased to work on Project Maven, the DoD contract was taken over by a company called Palantir.[52]

Drones use video imagery to recreate digitally, without physical risk, the human aerial observers of World War I. The drone pilots and video analysts are many miles away from their targets. While algorithmic recording of video information helps video analysts manage the hundreds of thousands of hours of content, it does not always assist in comprehension. But the immediacy of discovery and imprecise recording may not always lead to comprehension. Unlike previous wars that used surveillance for targeting, the targeting cycle has become so rapid that geospatial intelligence created by humans or algorithms with inadequate comprehension or incomplete context can cause inaccurate targeting of a misidentified individual.

As the volume of geospatial data races farther ahead of the human capacity to review it, much less analyze it, feature recognition technology has become essential for dealing with the volume of what satellites, aircraft, drones, and cell phones collect. Continuing developments in algorithmic recording, with feature recognition software on neural networks, have made impressive gains in improving the capability of the software to capture and record the known content of these images.

For humans to keep up with the amount of geospatial information, growth in the capability of these algorithms is essential. Yet, for certain discoveries from these sources of geospatial intelligence, the need for human attention remains constant. Since the Basic Encyclopedia computers have been accelerating the recording and information management of geospatial intelligence, yet the accelerating increases in collection and the corresponding decreases in analytic time and attention loom over the future of geospatial intelligence. Even when digital technology accelerates capturing metadata about the images, as well as measuring, identifying, and counting objects on the images, it remains true that the more that feature recognition technology discovers, comprehends, and records, the more that uniquely human capabilities become valuable for asking questions about discovery, comprehending, and tracking human activities in a world of denied and deceptive information.

While complex databases and GIS have automated and accelerated much of the recording, both technologies are bounded by the known. GIS requires an ontology, or classification structure, and a semantic structure, or agreed-upon meanings, to record accurately categories of geospatial information. But many indicators of new human activity initially are generic and ambiguous, and frequently too plentiful to document on the metadata of individual images. As an example, about 25 years ago, cell phone towers began to be built in rural landscapes, and while they were camouflaged they were viewed initially as technical intrusions on any natural landscape. Very quickly, they became too commonplace to notice.

During the Cold War, American imagery analysts had to search the Soviet Union for strategic infrastructure. With such a large area to look at, analysts became expert at identifying soviet military installations, often because of the incongruity of fence lines, new rail sidings in desolate areas, and improvements to roads in remote and uninhabited regions. That distinguishing feature, the fence line, would trigger attention that would be sustained on subsequent missions until the type of activity or installation could be identified.[53] Some installations could be identified well before completion by the configuration and number of fence lines. Recording metadata about these facilities was not challenging.

But many other common observations were more difficult to interpret. Often, analysts would observe only an area of scarred ground. If this scarred area was adjacent to a river or other body of water, a number of possibilities arise—some common, some uncommon, some benign, some malign. For instance, ground scarring on a river bank may indicate logging, agricultural clearing, pier construction, gravel dredging, if upstream from a town, a water intake plant; if downstream, a sewage treatment plant, or a future nuclear power plant. Until more human changes at the scarred area could be observed and categorized, imagery analysts had to keep an open mind and stay in the uncertainty of the question about all these possibilities and more.

When late in the Cold War, beginning in the mid-1980s, digital imagery and databases became common, the metadata challenge worsened rather than improved. The increased pace and volume of new geospatial data created a greater need for recording and resulted in many more incomplete observations. The proliferations of local data sets in imagery and geospatial organizations increasingly strained analysts and their information technology support. The exponential increase in geospatial data in the 21st century has multiplied these strains.

The thousands of commercial and governmental imaging satellites, the tens of thousands of imaging drones, and the billions of geospatially registered terrestrial imaging sensors in cell phones have resulted in an unknown number of separate repositories of geospatial data, much of which no eye has yet looked at and some of which no human will ever see. Even though essential metadata is now computer generated and recorded, this recording is currently limited to the sensor, the area of collection, and previously recognized objects.

While digital imagery enabled new kinds of recording, and the incorporation of GIS technology greatly accelerated the production and transmission of geospatial intelligence, they did not enlarge or improve analytic attention. When organizations focused the attention of their analysts, as the Western intelligence and cartographic organizations did during the Cold War, they have deep records about human changes on those parts of the planet. But the growth in volume of commercial imagery since the Cold war, and the proliferating number of small imagery satellites precludes a similar growth in human attention to other locations, so a recording challenge remains. The current number of orbiting imaging satellites can create more imagery than millions of humans can look at. Yet this new technology, in combination with neural network algorithms, can assist in a more rapid human recovery of knowledge and awareness of global security problems such as changes in climate, and shortages of water and food.

Most of the technological improvements in software in the 21st century have diminished the effort involved in geospatial recording, but it remains the necessary intellectual step after discovery and before comprehending.

NOTES

1 Finnegan, *Shooting the Front*, pp. 42, 162–165.
2 *Othello, The Moor of Venice*, Act III, Scene iii, line 360 *The Riverside Shakespeare* (Boston, MA: Houghton Mifflin, 1974), p.1224
3 Finnegan, *Shooting the Front*, pp. 46–7, Chapter 6, "The Operation of Interpretation," pp. 133–52. Babington-Smith, *Evidence in Camera*, pp. 20–24, 50–51, 80–82; Stanley, Roy, *V-Weapons Hunt*, pp. 38–40; Stewart, Paul, *Medmenham: Anglo-American Photographic Intelligence in the Second World War, Volume* 1, 55–58, Annex I, pp. 239–242, Brugioni, *Eyeball to Eyeball,* pp. 194–195, O'Connor, *NPIC,* pp. 46–47, 80–82, 111–112.
4 Babington-Smith, *Evidence-in-Camera*, pp. 59–60.
5 O'Connor, *NPIC: Seeing the Secrets, Growing the Leaders.* pp. 95–97.
6 Alastair Pearson documents the history of military model making in the 19th century in France and by England in World War I, but Paul Stewart's research indicates that the dissemination of hundreds of copies of individual models did not occur until World War II.
7 Stewart, Paul, Dr. *Medmenham: Anglo-American Photographic Intelligence in the Second World War, Volume* 1 (Submitted for the Doctor of Philosophy at the University of Northampton, 2019; Pearson, Alastair W. "Allied Military Model Making during World War II," *Cartography and Geographic Information Science*, Vol. 29, No. 3, 2002, pp. 227–241; Abrams, L.N. *Our Secret Little War* (Bethesda, MD: International Geographic Information Foundation, 1991). The use of military models has a long history, at least back to the 19th century in France. The first military use of a model based on

aerial photography was the Royal Navy raid on the submarine facilities in Zebrugge, Belgium, in World War I, but the first systemic combining of photography and model making was the British move of the model section to CIU in April 1941. Powys-Lybbe, Ursula. *The Eye of Intelligence* (London: William Kimber, 1983), p. 61.

8 https://en.wikipedia.org/wiki/Basic_Encyclopedia, viewed 23 June 2021.

9 Steven J. Collier and Andrew Lakoff. https://limn.it/articles/the-bombing-encyclopedia-of-the-world/, viewed 13 November 2021. Collier and Lakoff point out that the second Univac computer purchased by the US government in 1952 went into the Air Force Management Directorate which was responsible for the Basic Encyclopedia.

10 The National Reconnaissance Office's Center for the Study of National Reconnaissance has published declassified histories of the film return systems that chronicle the investments and the progress of these programs from their inception in 1958 through their conclusion in 1986.

11 www.cnas.org/publications/transcript/eric-schmidt-keynote-address-at-the-center-for-a-new-american-security-artificial-intelligence-and-global-security-summit.
www.usgs.gov/core-science-systems/nli/landsat/landsat-1?qt-science_support_page_related_con=0#qt-science_support_page_related_con, viewed 23 June 2021.

12 https://en.wikipedia.org/wiki/KH-11_Kennen, viewed 27 June 2021; *The Era Before KENNAN/KH-11, Section 2 KENNAN System.* p. 122 (Chantilly, VA. Center for the Study of National Reconnaissance, 2021). Approved for Release 2101/07/09 C05097636NRO.

13 Konecny, Gottfried. *Geoinformation: Remote Sensing, Photogrammetry, and Geographic Information Systems,* Second Edition (Boca Raton, FL: CRC Press, 2014), pp. 10–11.

14 https://history.state.gov/milestones/1969-1976/salt, viewed 15 November 2021. https://2009-2017.state.gov/t/avc/trty/102360.htm. Intermediate Range Nuclear Forces Treaty, viewed 15 November 2021.

15 https://media.defense.gov/1994/Nov/28/2001714972/-1/-1/1/95-043. pdf. Audit Report Office of the Inspector General, *Management of the Digital Production System Development at the Defense Mapping Agency*, Report Number 95-043, 29 November 1994; e-mail from author to Jack Hild, 19 April 2021.

16 O'Connor, *NPIC.* p. 184.

17 The governor on the pace of change was Moore's Law. Until its logarithmic effects on the speed of processing and the diminishing price of computer technology took effect in the computer industry, the vision of digital imagery lagged behind the reality of its implementation.

18 Rip, Michael Russell and James M. Hasik. *The Precision Revolution: GIS and the Future of Aerial Warfare* (Annapolis, MD: Naval Institute Press, 2002), pp. 60, 69,10, 92.

19 https://en.wikipedia.org/wiki/Jack_Dangermond, viewed 27 February 2021.

20 Kilday, *Never Lost Again*, Chapter 7.

21 www.theatlantic.com/international/archive/2011/08/inside-colin-powells-decision-to-declare-genocide-in-darfur/243560/, viewed 31 May 2021.

22 https://escweb.wr.usgs.gov/share/mooney/142.pdf. Overview of the 2010 Haiti Earthquake.

23 Zook, Matthew; Mark Graham; Taylor Shelton; and Sean Gorman (2010) "Volunteered Geographic Information and Crowdsourcing Disaster Relief: A Case Study of the Haitian Earthquake," *World Medical & Health Policy*, Vol. 2, No. 2, Article 2. DOI: 10.2202/1948-4682.1069. Available at: www.psocommons.org/wmhp/vol2/iss2/art2.

24 https://skytruth.org/2019/05/2017-frackfinder-update/,viewed 31 May 2021.

25 https://skytruth.org/2017/03/fracking-coming-to-a-backyard-near-you/, viewed 31 May 2017.

26 https://skytruth.org/2021/03/fighting-fracking-with-skytruth-alerts/, viewed 31 May 2021.

27 https://skytruth.org/2019/02/using-machine-learning-to-map-the-footprint-of-fracking-in-central-appalachia/, viewed 31 May 2021.

28 Higgins, Eliot. *We Are Bellingcat. Global Crime, Online Sleuths, and the Bold Future of News* (New York, NY: Bloomsbury, 2021), Chapter 2, "Becoming Bellingcat, A Team of Detectives Takes Shape," pp. 65–109.

29 www.youtube.com/watch?v=NkQ58I53mjk, viewed 23 June 2021. The skit dates from September 1952.

30 Smithsonian, https://americanhistory.si.edu/collections/search/object/nmah_334414, viewed 27 June 2021.

31 PIC/D-114/60,Memorandum for Deputy Director (Intelligence) Subject: Development of Automatad. 5 October 1960. SECRET, Declassified in Part: Sanitized Copy Approved for Release 2012/11/01: CIA RDP78B05702A000100070054-3.

32 In one of the last official acts of President Eisenhower, Lundahl's organization became the National Photographic Interpretation Center in January 1961. O'Connor, *NPIC*, p. 41.

33 Ecker, William B. and Kenneth V. Jack. *Blue Moon over Cuba: Aerial Reconnaissance during the Cuban Missile Crisis* (Oxford, UK: Osprey Publishing, 2012). The record keeping is imprecise about these missions. The high number is cited as 168, but the US Air Force and the US Navy

had separate record keeping systems, and the index to the mission numbers is not currently available.

34 O'Connor, Jack. "The Undercover Algorithm: A Secret Chapter in the Early History of Artificial Intelligence and Satellite Imagery," *International Journal of Intelligence and Counterintelligence*. DOI: 10 108008850607.2022.2073542

35 Metz, Cade. *Genius Makers: The Mavericks Who Brought AI to Google, Facebook and the World* (NY: Dutton, 2021) captures the history of neural-network algorithms and what creative and technological efforts were made to make them effective at recognizing features in imagery. The AlexNet won the ImageNet competition in 2012; Wei, Jerry. L viewed 23 June 2021.

36 Lohr, Steve, How Do You Vote: 50 Million Google Images Give a Clue," *New York Times*, 1 January 2018, Section B, Page 1.

37 www.fsa.usda.gov/programs-and-services/aerial-photography/imagery-programs/naip-imagery/, viewed 16 December 2021.

38 https://skytruth.org/2019/02/using-machine-learning-to-map-the-footprint-of-fracking-in-central-applachia/, viewed 1 June 2021.

39 Kanizsa, Gaetano. *Organization in Vision: Essays on Gestalt Perception* (New York, NY: Praeger, 1979), pp. 73–75, viewed 2 June 2021.

40 Humans who look at Kanizsa's image report seeing between zero and 11 triangles.

41 https://airandspace.si.edu/collection-objects/sa-2-guideline-missile/nasm_A19850424000, viewed 14 November 2021. The first observation of the missile by the west was at the 1957 Moscow May Day parade.

42 *Accomplishments of the U-2 Program*, 27 May 1960, p. 7, TCS-7519-60-b, Approved for Release 2001/03/30: CIA-RDP33-02415A000100070007-5.

43 www.jfklibrary.org/asset-viewer/archives/DODCMCBM/007/DODCMCBM-007-006My. colleague Joseph Caddell, Jr. points out that the pattern was so recognizable that the North Vietnamese, on receiving the SA-2 from the Soviet Union, altered the deployment pattern to increase their survivability.

44 Author's interview and e-mail exchange with Tom Maddox; Directorate of Intelligence, 2 May 1983, *SR-71 Mission over Cuba*, page 2, Sanitized Copy Approved for Release 2010/09/30: CIA-RDP85T00287R0004003600002-9. Cuba was one of the few foreign countries where the Soviet ruble was exchanged for local currency.

45 The history of these technical developments is well captured in *Cade Metz's Genius Makers: The renegades who brought AI to Google, Facebook, and the World* (New York, NY: Penguin Dutton, 2021) and Michael Wooldridge's *A Brief History of Artificial*

Intelligence: What It Is, Where We Are, and Where We Are Going (New York, NY: Flatiron, 2020).

46 Ursa.   www.ursaspace.com/blog/an-inside-look-at-sar-based-measu rements, viewed 27 June 2021.

47 https://ntrs.nasa.gov/api/citations/20170011719/downloads/2017 0011719.pdf Disaster Support Algorithms, viewed 27 June 2021.

48 Drew, Christopher. "U.S. Awash in Drones' Video Data," *New York Times*, 10 January 2010.

49 Drew, Christopher. "U.S. Awash in Drones' Video Data," *New York Times*, 10 January 2010.

50 Scharre, P., Cho, A., Allen, G., and Schmidt, E. 2017. Eric Schmidt Keynote Address at the Center for a New American Security Artificial Intelligence and Global Security Summit, 13 November. Accessed 18 January 2021.

51 The operational challenges in the war in Afghanistan of this form of recording are captured by Annie Jacobsen in *First Platoon: A Story of Modern War in the Age of Identity Dominance* (New York, NY: Dutton, 2021), Chapter 10, "The Gods-Eye View, pp. 124–140 addresses the challenges of providing accurate operational intelligence in this environment.

52 https://thenextweb.com/news/report-palantir-took-over-project-maven-the-military-ai-program-too-unethical-for-google.

53 Brugioni, Dino. A. "The Art and Science of Photoreconnaissance," *Scientific American,* Vol. 274, No. 3 (March 1996), pp. 78–85.

BIBLIOGRAPHY

Abrams, L.N. *Our Secret Little War* (Bethesda, MD: International Geographic Information Foundation, 1991).

https://airandspace.si.edu/collection-objects/sa-2-guideline-missile/nasm_A19850424000.

www.theatlantic.com/international/archive/2011/08/inside-colin-powe lls-decision-to-declare-genocide-in-darfur/.

Babington-Smith, Constance. *Evidence in Camera: The Story of Photographic Intelligence in the Second World War* (Phoenix Mill, UK: Sutton Publishing, 2004).

Brugioni, Dino A. "The Art and Science of Photoreconnaissance," *Scientific American,* Vol. 274, No. 3 (March 1996), pp. 78–85.

CIA PIC/D-114/60, Memorandum for Deputy Director (Intelligence) Subject: Development of Automatad. 5 October 1960. SECRET, Declassified in Part: Sanitized Copy Approved for Release 2012/11/01: CIA RDP78B05702A000100070054-3.s.

————*Accomplishments of the U-2 Program*, 27 May 1960, p.7, TCS-7519-60-b, Approved for Release 2001/03/30: CIA-RDP33-02415A000100070007-5.

————Directorate of Intelligence 2 May 1983, *SR-71 Mission over Cuba*, page 2, Sanitized Copy Approved for Release 2010/09/30: CIA-RDP85T00287R0004003600002-9.

Collier, Steven J. and Andrew Lakoff. https://limn.it/articles/the-bombing-encyclopedia-of-the-world/ viewed 13 November 2021.

Biography of a C-4 Stereoplanigraph. By: C. M. Cottrell and Milton Glicken, Fairchild Aerial Surveys. Reprinted from *Photogrammetric Engineering*, Vol. XXIX, No. 4, July 1963.

Drew, Christopher. "U.S. Awash in Drones' Video Data," *New York Times*, 10 January 2010.

Ecker, William B. and Kenneth V. Jack. *Blue Moon over Cuba: Aerial Reconnaissance during the Cuban Missile Crisis* (Oxford, UK: Osprey Publishing, 2012).

Eder, J. M. *History of Photography* (New York, NY: Colombia University Press, 1945).

https://escweb.wr.usgs.gov/share/mooney/142.pdf.

www.fsa.usda.gov/programs-and-services/aerial-photography/imagery-programs/naip-imagery/.

Higgins, Eliot. *We Are Bellingcat: Global Crime, Online Sleuths, and the Future of News* (New York, NY: Bloomsbury, 2021).

https://history.state.gov/milestones/1969-1776/salt1.

Jacobsen, Annie. *First Platoon: A Story of Modern War in the Age of Identity Dominance* (New York, NY: Dutton, 2021).

www.jfklibrary.org/asset-viewer/archives/DODCMCBM/007/DODCMCBM-007-006.

Kanizsa, Gaetano. *Organization in Vision: Essays on Gestalt Perception* (New York, NY: Praeger, 1979).

Kilday, Bill. *Never Lost Again: The Google Mapping Revolution That Sparked New Industries and Augmented Our Reality* (New York, New York: Harper Collins, 2018).

Konecny, Gottfried. *Geoinformation: Remote Sensing, Photogrammetry, and Geographic Information Systems*, Second Edition. (Boca Raton, FL: CRC Press, 2014).

Lohr, Steve. "How Do You Vote: 50 Million Google Images Give a Clue," *New York Times*, 1 January 2018, V. Section B, page 1.

www.mdshs.org/Biography%20of%20a%20C-4.html.

https://media.defense.gov/1994/Nov/28/2001714972/-1/-1/1/95-043.pdf. Audit Report Office of the Inspector General, *Management of the Digital Production System Development at the Defense Mapping Agency*, Report Number 95-043, 29 November 1994.

Metz, Cade. Genius Makers: *The Mavericks Who Brought AI to Google, Facebook, and the World* (New York, NY: Dutton, 2021).

NRO. Center for the Study of National Reconnaissance. *The Era before Kennan/KH-11.* www.nro.gov/Portals/65/documents/foia/declass/HISTORICALLY%20SIGNIFICANT%20DOCs/NRO%2060th%20Anniversary%20Docs/SC-2021-00002_C05097836.pdf. https://ntrs.nasa.gov/api/citations/20170011719/downloads/20170011719.pdf.

O'Connor, Jack. *NPIC: Seeing the Secrets, Growing the Leaders* (Alexandria, VA: Acumensa Press, 2015).

O'Connor, Jack. "Undercover Algorithm: A Secret Chapter in the Early History of Artificial Intelligence and Satellite Imagery," *International Journal of Intelligence and CounterIntelligence,* 2022. DOI: 10.1080/08850607.2022.2073542

Pearson, Alastair, W. "Allied Military Model Making During World War II," *Cartography and Geographic Information Science*, Vol. 29, No. 3, 2002, pp. 227–241.

Powys-Lybbe, Ursula. *The Eye of Intelligence* (London: William Kimber, 1983).

Rip, Michael Russell and James M. Hasik. *The Precision Revolution: GIS and the Future of Aerial Warfare* (Annapolis, MD: Naval Institute Press, 2008).

Scharre, P., Cho, A., Allen, G., and Schmidt, E. "Eric Schmidt Keynote Address at the Center for a New American Security Artificial Intelligence and Global Security Summit," 2017. 13 November. Accessed 18 January 2021.

Shakespeare, William, "*Othello, The Moor of Venice,*" in *The Riverside Shakespeare* (Boston, MA: Houghton Mifflin, 1974).

https://skytruth.org/2019/05/2017-frackfinder_update.

https://skytruth.org/2021/03/fighting_fracking_with_skytruth_alerts,

https://skytruth.org/2017/03/fracking_coming_to_a_backyard_near_you/.

https://skyrruth.org/2019/02/using_machine_learning_to_map_the_footprint_of_fracking_in_central_appalachia.

https://towardsdatascience.com/alexnet-the-architecture-that-challenged-cnns-e406d5297951#:~:text=AlexNet%20won%20the%202012%20ImageNet,labels%20on%20eight%20ImageNet%20images.

www.ursaspace.com/blog/an_inside_look_at_sar_based_measurements.

www.usgs.gov/core-science-systems/nli/landsat/landsat-1?qscience_support_page_related_con=0#qt-science_support_page_related_con.

www.whatispsychology.biz?Kanizsa_triangle_illusion_explanation.

https://en.wikipedia.org/wiki/Basic_Encyclopedia.

https://en.wikipedia.org/wiki/jack_Dangermond.

https://en.wikipedia.org/wiki/KH-11_Kennan.

Wooldridge, Michael. *A Brief History of Artificial Intelligence: What It Is, Where We Are, and Where We Are Going* (New York, NY: Flatiron Press, 2020).

www.youtube.com/watch?V=NkQ58153mjk (Lucille Ball).

Zook, Matthew, Mark Graham; Taylor Shelton, and Sean Gorman. "Volunteered Geographic Information and Crowdsourcing Disaster Relief: A Case Study of the Haitian Earthquake," *World Medical and Health Policy*, Vol. 2, No. 2, Article 2.

CHAPTER 5

Comprehending

INTRODUCTION

The fourth activity in geospatial intelligence is comprehending which requires more observations than discovery. Comprehending can also be characterized as a response to different questions. Discovery answers these questions—what, where, how many, and how large. Comprehending answers these questions—how and why, and sometimes, what's next. Comprehending also implies the knowledge of a process, sequence, or relation among multiple objects or locations.

An early example of comprehending dates back to World War I. It was the use of aerial photography to help determine the enemy disposition of forces and their infrastructure. Most of the elements of comprehending were initially developed in this conflict. Order-of-battle is the conceptual arrangement of military units, and aerial observation provided a new source for order-of-battle information. From observing the locations, specific numbers, and arrangement of men and equipment, the type and size of a military unit can be inferred. Repeated observations of infrastructure, a railroad yard, for example, enable a judgment about its routine volume of traffic, and changes in the routine can indicate a logistics buildup prior to an attack, withdrawal preparations, or offensive preparations before the movement of a military unit.

After World War I, aerial and space photography and imaging began to transform 20th century cartography. The growth in the use of stereo photography made the work of the interpreter much more precise. The ability of aircraft and, after 1960, satellites, to provide accurate, safe, and, most importantly, measurable access to areas denied by conflict, terrain, or cost, changed the nature of cartography. No longer was it necessary to have "eyes on" a particular terrain. Instead, technology increasingly provided a way to have "eyes on" film or digital images that could produce the same terrestrial measurements with similar or better

 DOI: 10.4324/9781003436836-6

accuracy. Improvements in mechanical creation of optics, aerial cameras, and stereo-comparators, in combination with the chemical improvements in film bases and chemical emulsions, greatly increased the accuracy and precision of mapping and imagery analysis. Increased precision became more important later in the 20th century.

This chapter discusses comprehending as a form of warning. By observing the German U-boat construction techniques carefully, the British photointerpreters in World War II were able to warn about upcoming launches. From aerial photography they could identify all the steps in the submarine construction process and estimate the rate of construction. The British also were able to determine the target orientation of the German fixed V-1 launch sites in Western Europe. This comprehension provided them with information about the future targets of these missiles as well as a way to attack them.

After the discovery of the Soviet missile test site at Tyuratam in August 1957, the analytic effort known as the Jam Session ensured that the US would develop aerial and space photographic intelligence that would provide comprehensive knowledge about Soviet strategic missile systems. Over the next 34 years, US study of the Soviet missile bases and silo construction techniques by photointerpreters and imagery analysts provided sufficient comprehension to provide the confidence for the US to enter arms control negotiations with the USSR with the certainty that space reconnaissance could detect potential cheating.

Comprehending one stage of a weapons development process does not always lead to comprehending the next stage. In World War II, the British comprehension of German development of the V-1 cruise missile and V-2 ballistic missile at the missile and aircraft test center at Peenemunde led to military actions, and the German responses to those actions made further comprehending much more difficult for British intelligence. This pattern of changing processes after recognizing that an enemy has comprehended what had been ongoing is common in intelligence. Knowledge gained from intelligence is always perishable.

Sometimes comprehending can take a long time. The Soviet Union experienced a nuclear disaster at Kysthym in 1957, but hid it successfully from the west for years. The initial aircraft and satellite information was not obtainable until 1960, and to enable discovery information from other intelligence sources had to be gotten long after the event. Only then after the site had been initially located in 1963 could the effort be initiated to comprehend the extent of the disaster.

An enduring challenge in intelligence, and geospatial intelligence, is distinguishing what is known from what is not known. The Cuban Missile Crisis, the best-known event involving aerial photography in the Cold War, demonstrates the risks in not sufficiently comprehending the full extent of the problem. The risks surrounding the deployed nuclear weapons during the Cuban Missile Crisis were not comprehended by the US or the Russians until 1993. Intelligence, even when it comes from multiple sources, rarely can achieve full comprehension. Warning without comprehending became a significant problem for US and Allied intelligence in Operation Iraqi Freedom in 2003. At the time US intelligence about Iraqi Weapon of Mass Destruction (WMD) had been based predominantly on imagery analysis, and the lack of other sources of intelligence about Iraqi WMD, and the insufficient imagery sampling to answer all the strategic questions, illustrate the risks of insufficient comprehending.

In contrast, the geospatial response a few years earlier to the unknown issues surrounding the onset of a new millennium, and the unknown status of global computer systems to accommodate the four-digit date change, illustrates the value of using a working hypothesis to assist in comprehending the future. The uncertainty about Y2K, as the turn of the millennium was then called, had to do with discovering if and where computer systems failed and the consequences of local failures. The global dependence on electrical power generation became the basis for the monitoring hypothesis. The US used a National Oceanic and Atmospheric Administration satellite to detect light emissions at night. This satellite source provided a means to tip-off the location of potential problems for further collection which thankfully turned out not to be necessary.

The final two examples discuss failures in comprehending: one, the 1998 Indian Nuclear Test, is the failure to recognize that once an adversary becomes aware of capabilities, it can modify its practices to defeat intelligence collection. The other failure results from the long-term lack of attention on a more subtle and challenging issue. The economy of the Soviet Union had been an intelligence question from the early 1950s through the 1980s. Yet, analytic efforts to answer economic questions throughout that time never had the same collection or analytic priority as questions about tactical and strategic weapons or regional conflicts. The study, or more appropriately the lack of geospatial study, of the Soviet economy illustrates the challenge of comprehending a slowly developing social change.

PHOTOGRAPHIC INTERPRETATION OF COMBAT BEGINS

During World War I, aerial surveillance and reconnaissance developed rapidly. For the first time since Leonardo da Vinci at Imola, humans were able to comprehend the military uses of terrain and spatial relations without having to traverse or survey the area of interest. The initial use of aerial photography was to gather more details about the immediate activity being surveilled and photographed. Throughout the war, the amount and extent of aerial photography grew among the French, Germans, and English armies. US efforts at aerial photography caught up with the other Allied nations in 1917 and 1918.[1] Much initial effort by all the forces went into accelerating the exploitation and reporting of information from the aerial photos. The value of the most recent images never diminished, and the demand for more and more aerial intelligence grew. The early interpreters and intelligence officers learned to look differently and better, and their efforts led to the first examples of comprehending.

THE ELEMENTS OF COMPREHENDING

A photographic image captures a moment in time, and World War I photointerpreters began to develop comprehensive knowledge of an area or a military unit by comparing the most current image with previous images of the same area. Comparison of images that captured moments in time enabled the interpreters to detect changes, and experience taught them to detect indications of future activities, as well as the meaning of certain changes. The static trench warfare that characterized much World War I combat meant that many locations were photographed repeatedly, even though the effects of artillery barrages may have changed the terrain significantly.[2] Detecting whether these terrain changes were a consequence of enemy shelling or intentionally created fighting positions became critical photo-intelligence. The photointerpreters learned that if they photographed areas with at least 60% overlap they could be examined with a stereoscope.[3] Stereo-photography provided information about the third dimension—depth and elevation—that was critical to analyzing terrain. Stereo collection also enabled the interpreters to see better into areas in shadow and to use shadow to distinguish between natural and man-made terrain features. Stereo photography also enabled more precise measurements of elevation.

Knowing the position of the camera as it took the photograph enabled photointerpreters to comprehend better. When pilots take a photograph that is not perpendicular to the target or at nadir (looking straight down), that photograph is called oblique.[4] While photography taken at nadir has higher resolution and less spatial distortion than an oblique image, sometimes an oblique photograph better displays patterns of shadow that can reveal details to the interpreter.

Another resource that helped interpreters in comprehending changes was photographic coverage over a larger area. This was obtainable in two ways, with different types of cameras or with multiple aircraft. Both the Germans and the Allies designed aerial cameras to collect larger areas on each image.[5] These cameras sacrificed higher resolution, or the ability to distinguish smaller objects on the ground, for the larger area of terrain covered in the field of view. The requirement to cover contiguous areas was also met by having multiple reconnaissance aircraft fly in close formation. This was riskier for the pilots and crews of these aircraft, but near the end of the war, when they had a larger number of aircraft to protect the reconnaissance aircraft from enemy fighters, the Allies had success with this tactic.[6] Coverage of contiguous areas over the trench lines and no man's land between opposing lines enabled situational reporting over a larger area. And as both sides developed their skills at comprehending, both began to fly and photograph further behind enemy lines.

Flights behind enemy lines allowed the photointerpreters to examine lines-of-communication, infrastructure, and enemy logistics. The military definition of a line-of-communication is the route that connects an operating unit with its supply base. Contiguous coverage behind the trench lines assisted analysts in comprehending the enemy's artillery dispositions, airfields, storage depots, and ammunition dumps. The photography also showed the building of temporary narrow-gauge railroads for logistics support from the rear areas to staging areas behind the trench lines. The study of the daily activity behind the front enabled the interpreters to establish a norm that would allow them to see and report atypical patterns and levels of traffic that sometimes indicated future attack preparations.[7]

By the end of World War I, aerial reconnaissance and defense against aerial reconnaissance had become a military necessity. Between 1915 and 1918 photointerpreters learned the elements of comprehension. Over that four-year period, military aerial photography had become an industrial process, and the number of reconnaissance missions increased throughout World War I. Both sides flew thousands of reconnaissance missions. The process of deriving information from aerial photographs was repeated hundreds of thousands of times, and all the combatant forces taught these processes in their service schools. In response to the growth in military photo-intelligence, after awareness of its principles became common knowledge, all combatants used measures to prevent enemy photointerpreters from discovering and comprehending their activities. As the World War I armies learned about the predictive value of aerial reconnaissance, they began using feints, and other deceptive measures, to conceal their intentions.

The first deceptive measure was camouflage. Combatants began to disguise equipment and positions by painting deceptive patterns on them. For experienced interpreters, camouflage sometimes became an indicator of something new or an object of greater importance.[8] Concealment was the second measure used by deployed forces to deceive aerial reconnaissance interpreters. Frequently units erected netting over the entrances to bunkers or installations, or over deployed artillery positions. Foliage was inserted into the netting to obscure what was beneath it. But the eyes of the interpreters often detected the variance between the camouflage and the background. Also, the lens and film could show variances in the tonal differences between live foliage in the scene and cut foliage in the netting.

All sides took deceptive measures when planning an attack or learning that they were under surveillance. They moved troops and equipment only at night or in inclement weather when aircraft could not fly. Units covered their tracks when they relocated their equipment. To divert attention from the area of an intended attack, deceptive artillery barrages and deceptive reconnaissance were conducted in other locations. To draw artillery and bombers away from actual targets, wooden artillery pieces and dummy aircraft were constructed.

In World War I photointerpretation was mostly tactical. The range of reconnaissance aircraft was no greater than fighter aircraft, and cameras could take only a limited number of photographs. A French pioneer of photointerpretation, Capitaine Paul-Louis Weiller, was among the first to understand that repeated study of the logistics targets behind enemy lines—rail networks, depots, ammunition dumps, airfields, and large-caliber artillery—could provide indications of future activity.[9] By the fall offensives in 1918, Weiller's thinking had spread to the US and British army intelligence units. Based on the additional analysis from aerial photographs behind the enemy lines, the Allies changed their assault tactics by shelling strategic targets at a greater distance behind the front lines. This tactic diminished the German ability to reinforce and resupply positions under attack.

Between 1915 and 1918, aerial photo-reconnaissance stopped being an experiment and became a military requirement. It had proven its value to commanders for current situational awareness, future operational planning, and, through the study of past photographs, capability assessments and indicators to enable warning of future enemy attacks. The industrialization of the processes of photo-reconnaissance and continuous process improvements brought about this transition.

The Allied forces employed the principles of scientific management, advocated and popularized by the American F.W. Taylor in the years before World War I, to refine the processes used to produce aerial photography in that war. From the onset of the war, all the armies pushed constantly to accelerate and standardize the photo-production process and cut the time between the landing of the aircraft and the delivering of developed and interpreted prints to the intelligence officers. The French, British, and American armies shared information and processes, and worked on standardizing camera types. Edward Steichen, the American photographer, was instrumental in these efforts. All the Allied nations made and shared constant innovations to minimize vibration, diminish the work the observer had to do in the air, and accelerate the chemical processing of plates and prints. While an aircraft could take between 20 and 40 photographs on a flight, sometimes thousands of copies of each of these prints were made within 24 hours.[10]

Later in the war, the number of print reproductions grew and, with the increasing number of daily missions, the photo labs made tens of thousands of prints. These intelligence efforts created information management, or recording, challenges for each of the military forces. But as rapidly as the forces learned how to collect, analyze, and disseminate all these aerial photographs, the November 1918 armistice stopped all these processes.

CARTOGRAPHY AND PRECISION

In wartime, intelligence organizations flourish, and in peacetime they are shrunk or disbanded. After World War I many of the lessons of aerial photo-reconnaissance were forgotten, and the tens of thousands of glass prints were destroyed to recover their silver content. The pace of developing technology to support this newest form of intelligence waned. Yet, after World War I, aerial photography would change cartography through the end of the 20th century, but the changes happened on the ground. The most significant post-war technical development occurred through the creation and development of stereo-plotters and comparators to measure the objects on these images. Between the wars, aerial photography businesses emerged in Europe and North America. This industry developed dedicated aircraft and aerial cameras that used film instead of glass plates to capture photographic images.

After World War I, several nations developed stereo-plotters designed to measure terrain from aerial photography for peacetime uses.[11] Germany, Austria, Switzerland, Italy, England, and South Africa all manufactured plotting instruments with the goal of improving the accuracy of distance measurements from aerial photographs. Academic work also continued on improving measurements from aerial photography. Of these developments, the most accurate stereo-comparators were produced by Zeiss and Wild, respectively, German and Swiss firms.

In the US and England, photographic interpretation and aerial reconnaissance efforts continued after World War I, but as exceptions rather than routine. The Royal Air Force flew aerial surveys in the British Empire over India, Afghanistan, Baluchistan (now part of Pakistan), Egypt and some photo-reconnaissance over Iraq in the early 1920s, but throughout the 1930s its resources to exploit aerial photography dwindled.[12] Prior to World War II, all the armies that would fight in that war had to reconstitute and enlarge their photointerpretation efforts.

Before World War II, the German military had designed unique aircraft for photo-reconnaissance, but their development of their photointerpretation and analytic processes was lacking.[13] The British developed no new technology, but as war with Germany became inevitable, a handful of individuals took analytic steps that made possible discovery and strategic comprehending by their intelligence services during World War II. Of particular note, British intelligence agencies wanted to obtain the most up-to-date European photogrammetric equipment. Before the declaration of war in September 1939, only two Swiss-manufactured Wild-stereo plotters existed in England. The British clandestinely procured Wild stereo-comparators in Europe and smuggled them into England. During the war, they used these cartographic tools to obtain precision mensuration on critical aircraft missions.[14] After December 1941, the US entered World War II and the British aided in the reconstruction of American aerial photography and photo interpretation efforts.[15]

World War I had taught intelligence officers the value of photographic context. For those who had difficulty seeing details on a photograph, the most easily perceived value was the currency of the image. More than any historical imagery, the military audiences for photointerpretation valued the most recent imagery over a specific area. For the photointerpreters, the challenges posed by incomplete recent coverage over a battlefield or current information that could not be negated or measured against a prior photograph, or current images full of difficult-to-discern details, particularly in shadow, constrained what they could tell intelligence officers, and what intelligence officers could tell military leaders and planners.

After the beginning of World War II, some World War I lessons had to be relearned quickly. The dynamic advances of the German military in 1940 and 1941 were much harder to follow and comprehend than World War I static trench warfare. After the British retreated from the continent at Dunkirk in May 1940, repeated aerial coverage of the German barge build-up in French ports in the summer of 1940 enabled the British to comprehend the scale and readiness of the German preparations for invading Britain. The British photointerpreters at the Central Interpretation Unit at Medmenham also noted the lengthening and enlarging of airfields in western France.[16] The photointerpreters' observations and the British military awareness of the amount of maritime support it would take to initiate and sustain a cross-channel military expedition enabled the comprehension and the warning about the German preparations.[17] At this point, less than a year into the war, the British had learned Sidney Cotton's lessons and flew these missions with fast, high altitude photo-reconnaissance Spitfires.

In their analysis of German submarine (U-boat) construction throughout the war, the British photointerpreters again demonstrated the utility of periodic coverage to achieve comprehension. German ship construction then took months. Throughout the war, repeated British overflights of the U-boat construction yards at Kiel and Hamburg provided an index of the rate of U-boat construction.[18] This index enabled accurate estimates throughout the war about the rate of growth and replacement in the German submarine fleets. This comprehension was achievable only because the British flew reconnaissance missions, with considerable risk to the pilots, at a frequency greater than the rate of construction. The essence of aerial photo, imagery, or geospatial collection management is sampling the target at a rate faster than the process or activity that is being imaged. Insufficient sampling precludes comprehending.

The distinction between discovering and comprehending can be marked by a change in the question the analyst is trying to answer. Discovery, which has its basis in photointerpretation, ends at the moment when the questions, *where, how long*, and, sometimes, *what*, can be answered. Comprehending begins as the questions *how, why*, and *"in relation to what?"* emerge, and the answers to these questions require imagery analysis.

The movement from photointerpretation questions to imagery analysis questions requires some intellectual attributes not always needed for discovery, as it is a movement from answering empirical questions—those requiring seeing in the present—to answering intellectual questions—those requiring thinking about the future. One of the earliest and most famous examples of photographic comprehending emerged in World War II. The British photointerpreters' analysis of German aerial and missile weapons development at Peenemunde illustrates the differences between discovery and comprehending.

Constance Babington-Smith's account of discovering tail-less aircraft, and Andre Kenny's and Hamshaw Thomas's observations of unidentified tower-like structures, other structures that looked like ski ramps, and sausage-shaped unidentifiable objects at Peenemunde provide excellent examples of what now be would be called geospatial discovery.[19] More importantly, Babington-Smith's and other accounts of the subsequent analysis that followed these discoveries illustrate the importance of research, in particular, repeatedly looking again at older imagery as part of the process of comprehending.[20] After her initial discoveries, Babington-Smith sought additional collection over Peenemunde, the research facility on the German coast of the Baltic sea where the discoveries had been made. But her next step became equally important, as she coordinated her preliminary findings among other British photointerpreters and intelligence analysts working air force issues.

THE PEENEMUNDE DISCOVERIES AND PRECISION MEASUREMENT

Babington-Smith's communication with her leadership, particularly Douglas Kendall, marked an important second analytic step. She followed up by making additional inquiries outside her organization. She reached out to R.V. Jones, a science advisor to Prime Minister Churchill, whom she had briefed previously. Jones had access to other intelligence sources, notably human intelligence reports and ULTRA communications intelligence intercepts, the most secret of the British World War II intelligence sources. Unlike academia, where research begins with what is known and provable, intelligence research often begins with what is suspected, and that may or may not be provable. Although not cleared for these other sources of information, Babington-Smith suspected the existence of other sources. R.V. Jones encouraged Babington-Smith to continue her inquiries. He had information from human sources who had reported unusual activities and very high security measures at Peenemunde, and Jones also engaged in his own private photo interpretation.[21]

With Jones's assistance, Babington-Smith helped identify two German weapons development programs that would become significant strategic challenges for Britain in the European theater in World War II. With the assistance of her Medmenham colleagues, Babington-Smith identified the Arado jet fighter program (Komet), and the V-1 cruise missile program. At the same location, Andre Kenny and Hamshaw Thomas identified the V-2 ballistic missile program.[22] Comprehending the 1942 and 1943 observations of developments at Peenemunde, and determining the characteristics and functions of all three new and unique weapons systems, took analytic time, attention, and resources.[23]

After the British government had sufficient information to recognize the potential threat from the ongoing research and testing at Peenemunde, it commissioned a large bombing raid in August 1943. While the raid slowed the testing program, it alerted the Germans to British awareness and intentions. Germany began constructing strategic weapons production facilities far away from Peenemunde in difficult-to-locate and underground facilities.

INFERENTIAL COMPREHENDING OF
V-WEAPONS LAUNCH SITES

While photographic study of Peenemunde led to comprehending the types of weapons under development there, once these weapons were deployed, only partial comprehension followed. At Peenemunde, British photointerpreters identified three fixed launch sites along the waterfront and gradually understood that they were a signature for the initial V-1 deployments.[24] After construction of these "ski-ramp" launch sites began to be detected in France by the British and American photo-interpreters at the Allied Central Interpretation Unit at Medmenham, the Allies bombed these sites. In turn, the Germans began to construct portable and temporary launch sites which Allied photointerpreters had a much more difficult time finding and identifying. While Allied photointerpreters began comprehending the nature of the V-1's operations at the Peenemunde test site, they were unable to gain enough comprehension to counter the deployment of these weapons.

V-1 launch sites had been easier to detect, but not easy to comprehend on the shoreline at Peenemunde in 1943. Even after their purpose had been comprehended, they were much harder to detect in the forested, rural, and village areas of Western Europe in the winter of 1943 and the spring of 1944.[25] Reconnaissance pilots flew hundreds of aircraft sorties and photointerpreters searched the film intently for these launch sites. Initially, the photointerpreters succeeded at identifying the permanent sites, and with the aid of the photogrammetry obtained by using the Wild stereo-comparator, they created a map that identified, on the basis of the launch azimuths, the English targets—Plymouth, Southampton, and London—for these new weapons (Figure 5.1).[26]

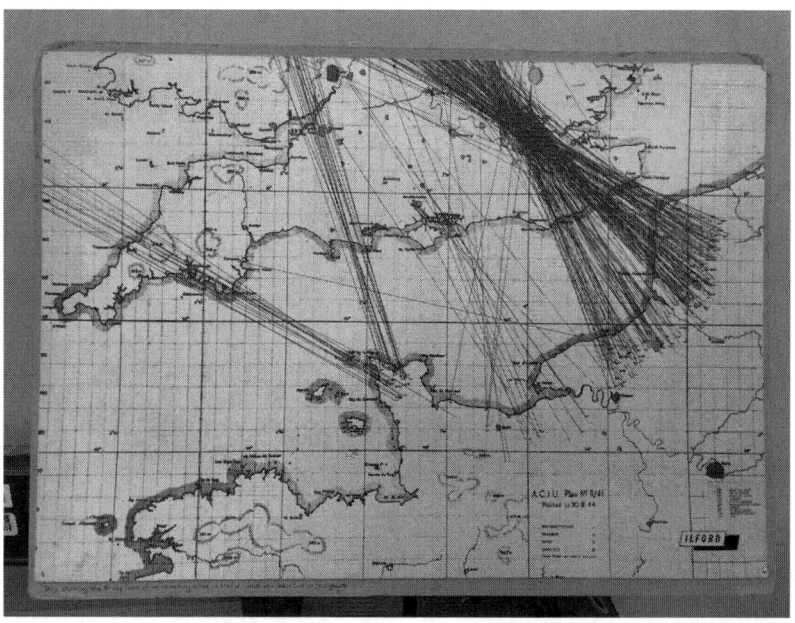

FIGURE 5.1   ACIU Map showing azimuths of German V-1 launch sites in August 1944.

Source: Courtesy of the Medmenham Collection.

The Allied comprehending of the infrastructure and order of battle for the V-1 resulted only in limited success in countering this weapon. The pace of the Allied Advance in western Europe diminished much of the German missile threat, but V-1 and V-2 attacks continued until April 1945, almost the end of the European war.[27]

COLD WAR COMPREHENSION

After World War II when the nuclear age began, shrinking the time between discovering and comprehending strategic weapons systems became a US intelligence priority. From 1948 through 1956, the early years of the Cold War, neither the US nor any other nation knew much about the Soviet Union, much less its secret nuclear and missile-related weapons. The partial extent of geographic knowledge of the Soviet Union had mostly come at the end of World War II from a daring American military intelligence effort by a handful of men and women led by Floyd Hough, a former surveyor, engineer, and geodesist.[28]

The Hough team brought an index of Army Map Service data to Europe. Their mission was to obtain[29] from recently reconquered areas as much captured German and European geographic and geodesic data as possible. Although a small group, they successfully recovered a great deal of survey points information, including survey points of parts of Germany that the Nazis were still defending. These were immediately rushed to American artillery units. When the Hough team went to Saalfeld in southern Germany, they found the central map depot and geodetic data depository for the German Army. This warehouse held mapping and geodetic information about eastern Europe and the Soviet Union. Also, the Hough team found seven Zeiss, state-of-the-art planigraphs that were taken to the US and put to cartographic use.[30]

After the retrospective discoveries of Soviet atomic and hydrogen weapons in 1948 and 1952, the US recognized that it faced much greater risks on account of the scarcity of its information about the Soviet Union. This recognition magnified the importance of Hough's efforts. The lack of geographic and geodesic awareness of the Soviet Union east of the Urals where the German army could not penetrate precluded obtaining basic mapping information of that region. Apart from Hough's information from Germany in 1945, no more recent cartographic or geodesic information from the USSR and the Warsaw Pact countries was available. Comprehending the geography and then accomplishing the strategic mapping of the Soviet Union would be a Cold War challenge for many years. In 1956, Western ignorance about the Soviet Union was defined starkly in the US Intelligence Community, when in the initial planning for U-2 overflights, the US military and intelligence community could identify only 45 strategic locations in the entire Soviet Union for which they had latitudes and longitudes.[31]

In 1956, the U-2 program, the U.S. Intelligence Community response to diminish this risk, began overflight operations. It brought many early discoveries about the Soviet Union, as well as greater awareness of how little of the USSR has actually been photographed. Comprehension about Soviet strategic weapons did not begin until after August 1957 with the onset of the Jam Session. Between 1956 and 1960, the US took the risks to fly deeply and clandestinely into the Soviet Union 28 times to obtain information about the Soviet nuclear and strategic weapons infrastructure.

INCREMENTAL COMPREHENDING

The development of high-resolution photo satellites in the early 1960s provided the means to comprehend the Soviet missile silos being constructed along the Soviet railnet. Starting in 1963, the use of the high-resolution KH-7 and KH-8 GAMBIT satellites—satellites with cameras that could provide photography as detailed as the U-2 aircraft cameras—provided photointerpreters with multiple high-resolution observations of silo construction. Yet this technology would have been of little use without the repeated attention and recording of the observations of individual targets and suspected areas associated with Soviet surface-to-surface missile systems.

Over a decade (1959–1969) the National Photographic Interpretation Center (NPIC) wrote more than 550 reports on silo construction in the Soviet Union.[32] With the help of photogrammetrists, two analysts at NPIC determined the Soviet silo construction sequences and chronology by measuring the shape and size of the individual components used in SS-9 and SS-11 silo construction. In the late 1960s, these two analysts wrote a report and from their study of the components the NPIC model shop created a three-dimensional model. The report found a wide audience in the US arms control community and the model was used in more than 120 briefings.[33] The two analysts and the photogrammetrists who measured the SS-9 and SS-11 silo components would have failed at their analysis had the photogrammetrist not had access to a computer to assist with the mensuration, and if the analysts had been unable to have their repeated observations and the photogrammetrist's repeated measurements recorded into a database.

Their report and the model convinced the US arms control community that NPIC, using satellite imagery, or national technical means, as it was described in the unclassified summary of the negotiations, could detect any attempt to cheat on the terms of the SALT treaty. The analytic effort that began with the efforts of these two analysts created the underpinning that provided confidence for US Arms Control negotiators during the first Strategic Arms Limitation Talks with the Soviet Union in 1969–1979.[34] As importantly, the ability of the two analysts to answer accurately the questions about how the Soviet Union constructed silos changed the perception in the intelligence community about the skills of imagery analysts. After their work, the intelligence community began to incorporate imagery more frequently in its responses to other intelligence questions.

DELAYED COMPREHENDING

Comprehension, often sought in geospatial Intelligence, can take years to achieve. While the Jam Session was being planned, about 900 miles north of Tyuratam an accident occurred near Kysthym on 29 September 1957. At a nuclear storage facility called Mayak, a cooling system for radioactive waste failed. This failure caused the stored liquid nuclear waste material to heat up. When it reached critical mass, it exploded with enough force to blast a 160-ton concrete lid off the storage area, and to release a radioactive airborne plume sufficient to irradiate and contaminate an area as large as Massachusetts or Haiti.

On account of the Soviet secrecy about their nuclear infrastructure and the scarcity of the U-2 collection at the time, the exact location of the Kysthym explosion was not known until 1968.[35] The CORONA satellite program did not successfully photograph Kysthym until 1963 but this satellite lacked the resolution to precisely identify the location of this accident. Not until the higher resolution KH-7 was on orbit could some of the evidence of the accident, the cordoned-off area, be identified. The details about the cause of the event did not emerge until the disclosures of Zhores Medvedev in 1976 and 1980.[36]

THE CUBAN MISSILE CRISIS AND INCOMPLETE COMPREHENDING

The U-2 program initiated the decades-long effort that enabled comprehension of the Soviet missile programs. While it was very successful at discovery in Cuba in October 1962, it did not achieve the same level of comprehending. Very quickly, the discovery of the Soviet intercontinental ballistic missiles (ICBM) and intermediate-range ballistic missiles in October achieved the status of an intelligence myth in the US. The successful part of the story has been told repeatedly, starting with President Kennedy's briefing to the nation on 22 October 1962; Secretary McNamara's presentation with John Hughes on 6 February 1963,[37] and several books on the subject by a number of participants, among whom were Robert Kennedy, Daniel Ellsberg, and Dino Brugioni. Yet on the 30th anniversary, after the fall of the USSR, disclosures by US, Russian, and Cuban participants and subsequent scholarship indicate that there was a significant lack of comprehension at the time of the threat by the US Intelligence Community and the Soviet Union. The 1992 post-Cold War meeting of the participants revealed[38] the extent of US ignorance of the number and location of the short-range nuclear tipped missiles, the warheads for the SS-4 and SS-5 missiles, and the nuclear-armed Soviet submarines. The mutual ignorance on both the US and Soviet Union of some of the most lethal weapons illustrates the risks of partial comprehension.

Cuba was the first crisis in which imagery and geospatial intelligence succeeded in discovering and failed in comprehending the full extent of the risk posed to the US and the Soviet Union by the deployment of nuclear weapons in Cuba. Forty years later, in a different crisis, the US would repeat the experience of limited comprehension of weapons of mass destruction (WMD), with serious but thankfully not catastrophic implications.

IRAQ: A LACK OF COMPREHENDING

By 2002, in the preparations for the second US–Iraq war, imagery and geospatial intelligence had been employed to observe and map Iraq frequently since 1990. Much of this imagery collection supported Operations Northern Watch and Southern Watch, which monitored potential Iraqi military attacks on the Kurds in northern Iraq and on the Shia in southern Iraq. Yet all this coverage and attention to Iraq suffered from the same comprehension challenge as some of the 1962 analysis of Cuba.

The volume of imagery intelligence on Iraq in 2002 far exceeded the intelligence from all other sources combined.[39] Following the invasion of Kuwait in August 1990, through Desert Shield/Desert Storm and U-2 imagery support to the UN inspectors from 1991 through 1998, U-2 aircraft missions were continually flown over northern and southern Iraq to protect the Kurds and Shia, respectively.[40] After the UN WMD inspection teams withdrew for intervals in 1998 and 2003,[41] nearly all subsequent information about potential WMD in Iraq came from aircraft and satellite imagery.

In general, when the preponderance of information comes from a single intelligence source, skepticism about the analysis and qualification about the reporting ought to increase. The failure of US intelligence to find evidence of Iraqi WMD programs, a significant part of the rationale for invading Iraq, illustrates the risk of not comprehending. As no intelligence source can answer all intelligence questions, successful comprehending, particularly geospatial comprehending, requires more than a single intelligence source.

While imagery or geospatial intelligence is very good at answering questions about capability and spatial relations, specifically, what, where, how many, and sometimes how and why, it struggles with temporal relations except in times of synoptic surveillance. Geospatial intelligence, without other sources, cannot answer intelligence questions about motivation and intent, particularly in locations where camouflage, concealment, and deception continue to be practiced and refined. Research, consultation, and collection are not the only requirements for comprehending geospatial intelligence issues. For some intelligence problems, a model can accelerate comprehension.

Y2K: THE CRISIS THAT DID NOT HAPPEN

In 1999, the US Intelligence Community planned to monitor the potential global effects of Y2K. Y2K was the shorthand expression for the Julian calendar change to the new millennium, in which the first two dates of the following year would begin with 20 instead of 19. At the time the fear was that a large amount of existing computer coding could fail as the earlier 20th century programmers working under storage constraints had not allowed for the change in the first two digits of the number indicating the new year. In 1999, the expectation was that some computers would not know what year it was and malfunction. The discovery challenge was global, the event was considered singular and potentially unique, and the global nature of the potential problem made it an intelligence issue.

As the US Intelligence Community planned how to collect information to locate and analyze this potential problem, it proceeded on the assumption that comprehending the potential issue would depend on analysis of installations that required electricity to function safely. These installations included nuclear energy plants, nuclear waste storage facilities, hospitals, water treatment plants, and food processing facilities.

Thankfully, Y2K turned out to be much ado about nothing. It was a classic example of negative intelligence, in which the absence of Y2K-caused computer issues turned out to be a global good. No significant power outages or large-scale events were caused that day by computer failures. But Y2K does illustrate the importance of using a model—electric power generation—to illuminate second-order questions in advance so that upon discovery, geospatial intelligence organizations could focus on comprehending the potential effects of the issue.[42] The US Air Force

Defense Meteorological Satellite System, which has the ability to detect light emissions at night, was essential to this effort.[43]

INDIA: DENIED COMPREHENDING

Even with research, target familiarity, sufficient collection, and a model, geospatial intelligence can fail to comprehend. In May 1998, at a known Indian underground test site in the Thar desert, the Indian government successfully conducted a nuclear test. While the US Intelligence Community had known the location of this test site for more than 20 years, comprehended how the Indian nuclear organization tested nuclear weapons, and had tracked developments at Thar desert, it still failed to provide any warning about this event.

Multiple reasons exist but do not fully explain this failure. Because the US had studied Indian nuclear test practices after the first surprise nuclear test in 1974, it assumed that the visible indicators of an upcoming nuclear test would remain the same. This assumption was made even though the US had interceded with the Indian government to prevent a prior nuclear test. In 1990, the Bush administration sent an envoy to India with satellite images and a demarche about their nuclear test preparations. The Indian government, at that time, chose not to continue the nuclear test, and it also learned about US imagery capabilities from the experience.[44]

In December 1995 the US again provided information to the Indian government that revealed how it monitored their test preparations.[45] One of the challenges in geospatial intelligence, with antecedents back to World War I, occurs when an adversary learns how an opponent looks or surveils. Subsequently, the adversary can introduce camouflage or concealment to deceive the opponent and to deny the expected observations.

In effect, the US government taught the Indian government how to avoid detection of its nuclear tests by US satellites. India, which has had its own space program since 1962, had the technological and orbital expertise from its own space program to analyze exactly when the US imagery satellites would be able to observe its test site. India also had the operational expertise to alert its workers to avoid the daily intervals when their surface activities would be subject to observation from space. The combination of US revelations and Indian competence may have helped cause this intelligence failure. The Indian nuclear test illustrates that comprehending becomes much more difficult, if not impossible, when the opponent knows the imagery sampling schedule, location of interest, and satellite resolution capabilities. The 1998 Thar desert test illustrates why geospatial intelligence officers must always be suspicious that deception or denial efforts may be ongoing.

Two opposite characteristics conclude this discussion of comprehending—the disciplined recording of information in a database, and the skepticism to ignore the received wisdom about the past recordings. Disciplined recording of observations began in World War II and continued through the Cold War. Without the disciplined recording early in the Cold War of the initial U-2 penetration missions and the analysis of the Jam Session, the collection strategy and subsequent comprehension that increased the Soviet database from 45 to more than 10,000 national targets could not have been accomplished. The important byproduct of that accomplishment, the US confidence to enter into arms control treaty negotiations with the USSR from the 1970s through the 1990s had its basis in the hundreds of reports and thousands of database entries before the successful negotiations in which satellite reconnaissance became the agreed means of verification.

And yet this achievement, accomplished over 40 years with imprecisely derived coordinates, limited collection sampling, insufficient resolution, adequate swath width for context, un-synchronized databases, and geospatially registered visual reporting (shape files), also points out what cannot be done. For areas or issues that are not the subject of focus and attention, comprehension is not possible.

PRECLUDING COMPREHENDING

The fall of the Soviet Union and the Warsaw Pact governments in the late 1980s was rapid and a surprise to western intelligence agencies. Nearly all the prior intelligence and geospatial discoveries and comprehension did not help to provide warning about these political and social changes. The nearly 40-year effort to attain comprehension about the military forces and strategic infrastructure of the Soviet Union and its allies did not extend to the civil institutions like the domestic economy, the social order, and the Soviet governmental ability to deliver social services.

When government organizations focus on one set of priorities, as the US Intelligence Community did with its National Tasking Plan during the Cold War, they also identify the issues where they lack focus. For all the attention the US and European intelligence agencies placed on USSR strategic and military forces, they equally did not focus on Soviet economic and social issues. In Richard Bissell's reminiscences, written after the fall of the Soviet Union, and published two years after his death in 1994, he quotes James Reber about this issue.

Bissell was responsible for two of the CIA's greatest successes—the U-2 program and the first photo-satellite, CORONA. He was also responsible for one of its greatest failures—the Bay of Pigs invasion of Cuba in

1961. Reber had been responsible for prioritizing intelligence collection and identifying targets for the U-2 and the early photo-satellite programs.

> Recently Reber has raised some good points about how the cold-war mentality caused us to overlook industrial and agricultural targets that might have helped us better evaluate the Soviet Union's true economic capacity and anticipated its decline. Locked into a mindset that focused on military targets, we didn't raise the right questions. From his perspective, the entire intelligence community should have developed the requirements needed to evaluate the economic stability of the Soviet system. Had we focused on such targets as "all of those physical photographable elements of a transportation system," we would have learned that the Soviets were putting little money into the nonmilitary productive resources that truly make a nation strong.[46]

The surprising pace of events in 1991 and 1992 with the collapse of the Soviet Union and the Warsaw Pact spotlighted this gap in the National Tasking Plan that had governed US imagery intelligence collection since 1975. This failure does not only apply to geospatial intelligence, but also to the entire US Intelligence Community. In any level-of-effort program[47] there will be intelligence questions that do not meet the level of effort. A challenge for individual geospatial analysts remains to consider the questions not being asked about the subject of focus and to reflect regularly on what is not the current subject of focus.

Comprehending requires more frequent collection sampling of a target than discovery, and tracking, the final intellectual activity of geospatial intelligence, requires more frequent sampling than comprehending.

NOTES

1 Finnegan *Shooting the Front.*
2 Finnegan. P. 305–316.
3 This technology emerged in the 1850s and was refined up to the war's beginning. (Eder, J.M. *History of Photography* (New York, NY: Colombia UP, 1945), pp. 381–384.
4 Finnegan, pp. 124–127.
5 Finnegan, pp. 135–138.
6 Finnegan, pp. 340–342.
7 Finnegan, pp. 119–120, 261–272. The French developed strategic research, then called strategical research, and late in the war, Capitaine Paul-Louis Weiller led three squadrons of reconnaissance aircraft dedicated to this mission.

8 Finnegan. pp. 317–331.

9 Finnegan, pp. 261–269.

10 Hardesty, Von. *Camera Aloft: Edward Steichen in the Great War* (New York, NY: Cambridge University Press, 2015), pp. 171, 184; Finnegan, p. 113.

11 Petrie, G. "A Short History of British Stereoplotting Instrument Design," *Photogrammetric Record*, 9 (50), 213–238 (October 1977), pp. 214–216, 234; Burtch, R. Class Notes Sure 340, 2009, *History of Photogrammetry*, Ferris State University, pp. 25–27.

12 Conyers-Nesbitt, Roy. *Eyes of the RAF: A History of Photo-Reconnaissance* (Phoenix Mill, UK: Sutton Publications, 1996), pp. 54, 58.

13 Stanley, Roy M. *World War II Photo Intelligence* (New York, NY: Scribners, 1981), pp. 116–127. The comment on their processes is on pp. 125–126.

14 Williams, Allen. *Operation Crossbow: The Untold Story of the Search for Hitler's Secret Weapons* (London: Random House, 2013), pp. 98–100; also, Ruth Pooley and Nick Perry, "Wild War Stories," *The Medmenham Magazine*, Spring 2021, pp. 26–27, and "Wild War II," *The Medmenham Magazine*, Autumn 2021, pp. 38–41).

15 The United States, at this time, had only one Zeiss stereo-comparator, owned by Fairchild Aerial Surveys, "Biography of a C-4 Stereoplanigraph," C. M. Cottrell and Milton Glicken. www.mdshs. org/Biography%20of%20a%20C-4.html, viewed 14 December 2021.

16 Stewart, p. 78.

17 Babington-Smith, *Evidence in Camera*, pp. 60–63.

18 Babington-Smith, *Evidence in Camera*, pp. 94–98.

19 Babington-Smith, *Evidence in Camera: The Story of Photographic Intelligence in the Second World War* (Phoenix Mill, Gloucestershire: Sutton Publishing, 2004), pp. 173–184.

20 See also Ursule Powys-Lybbe, *The Eye of Intelligence*, Roy Conyers Nesbitt, and R. V. Jones.

21 Jones, R.V. *Most Secret War* (London: Hamish Hamilton, 1978), pp. 338.

22 The controversy over this discovery is captured in the contrasting narratives of Jones in *Most Secret War*, and Powys-Lybbe in *The Eye of Intelligence*. The multiple narrative versions are noteworthy in that each describes differently the extent of analytic ambiguity and disagreement.

23 Babington-Smith, Evidence in Camera, pp. 176–208; Ursula Powys-Lybbe, *The Eye of Intelligence*, pp. 188–208.

24 Stanley, Roy. M. *V Weapons Hunt: Defeating German Secret Weapons* (Barnsley: Pen and Sword, 2010), pp. 60, 81–84; Powys-Lybbe,

Ursula. *The Eye of Intelligence* (London, William Kimber, 1983), pp. 202–203.

25 Powys-Lybbe, *The Eye of Intelligence*, pp. 195–196; Stanley, *V Weapons Hunt*, pp. 43.

26 ACIU Map showing azimuths of German V-1 launch sites in August 1944. (The Medmenham Collection).

27 My former colleague, Joseph Caddell, Jr. considers that the ACIS effort against the V-weapon deployment could be considered the earliest example of geospatial intelligence. I respectfully disagree as there was not an organized attempt in the English government to bring experts from other intelligence disciplines to partner with the photointerpreters. The Sandys committee had been organized to coordinate all the intelligence, but it did not pull together analysts from different disciplines to work side-by-side. The individual efforts of R. V. Jones at times seem to cooperate with and at other times to compete with the ACIS photointerpreters (Williams, *Operation Crossbow*, p. 309). The ACIS organizational structure allowed only the commander, Douglas Kendall, access to ULTRA information. This limited, for security reasons, the effectiveness of the analysis done by photointerpreters. Finally, the ACIS discoveries, while captured on an extraordinary example of spatial analysis, did not influence future cartographic efforts against this issue or Germany.

28 Geodeists study the size and shape of the earth.

29 I first learned about the HOUGH team from the National Archives research efforts of Thom Kaye, with whom it was my fortune to work in the short-lived Center for the Study of Geospatial Intelligence. These discoveries were first published on his blog, http://disturbedgeograp her.com. The story became more widely known with Greg Miller's article, "Behind the Lines: The Untold Story of the Secret U.S. Mission to capture priceless Mapping Data Held by the Nazis," *Smithsonian Magazine*, November 2019, pp. 64–78.

30 Planigraphs were used for the stereo plotting of aerial photographs to accurately measure elevation data. Prior to Hough's discovery, only one Zeiss stereoplanigraph existed in the United States, owned by Fairchild Aerial Surveys, "Biography of a C-4 Stereoplanigraph," C. M. Cottrell and Milton Glicken. www.mdshs.org/Biography%20 of%20a%20C-4.html, viewed 14 December 2021.

31 Ad Hoc Requirements Committee on Project AQUATONE (ARC) Minutes of Meetings Held in Room 121 Administration Building, Central Intelligence Agency, at 1200, 1 June 1956 and at 1400, 4 June 1956. SAFC 678 Approved for Release 2000/08/21/ RIA-RDP33-02415A000100100056-7).

32 More than 60-years later, the KH-7 and KH-8 capabilities have not been declassified. The number of NPIC reports is derived from the

author's queries for reports published in each of these years with the title "Soviet Silo Construction NPIC," in www.cia.gov/read ingroom/, the repository for declassified CIA reporting (viewed 13 December 2021). Each of these reports would have had their genesis in uncounted data base entries, and mission summaries. The queries included mission summary reports (OAKs) and indexes of photographic reports that were not counted as were the obvious errors—reports on another country or on another weapon or strategic system.

33 O'Connor, Jack. NPIC: *Seeing the Secrets and Growing the Leaders* (Alexandria, VA: Acumensa, 2015), p. 69; also Brugioni, *Eyes in the Sky*, pp. 395–396.

34 Newhouse, John. *Cold Dawn: The Definitive History of SALT* (New York, NY: Reinhart and Winston, 1973), p. 70; Wikipedia, SALT-I and SALT-II; https://history.state.gov/milestones/1969-1976/salt, viewed 25 June 2021.

35 CIA. *U.S.S.R.: Nuclear Accident Near Kysthym in 1957–58*, An Intelligence Assessment. National Foreign Assessment Center, SECRET, SW81-10102, October 1981. CIA Historical Review Program Release as Sanitized 1999. Kysthym was one of the targets for collection on the final U-2 penetration mission over the U.S.S.R. flown by Francis Gary Powers that was shot down on 1 May 1960. (Brugioni, *Eyes in the Sky*, p. 346). The public awareness of the accident followed the revelations of a Soviet Dissident, Zhores Medvedev (Medvedev, Zhores. A. Nuclear Disaster in the Urals. Tr. George Saunders (London: Angus and Robertson, 1979, Chapter 3, pp. 19–27).

36 Richelson, *Spying on the Bomb*. p. 129.

37 Brugioni, *Eyeball to Eyeball*, p. 210.

38 Naftali, Timothy, and Aleksandr Fursenko. *One Hell of a Gamble; Khrushchev, Castro, and Kennedy 1858–1964.* (New York, NY: W.W. Norton & Co., 1997), pp. 188, 211–12; Dobbs, Michael. "Tracing the Nuclear Warheads, One Minute to Midnight: Kennedy, Khrushchev, and Castro on the Brink of Nuclear War." Posted 18 June 2008. https://nsarchive2.gwu.edu/nsa/cuba_mis_cri/dobbs/warheads.htm.
Viewed 10 March 2021; ed. Svetlana Savranskaya and Thomas Blanton with Anna Melyakova. "Last Nuclear Weapons Left Cuba in December 1962. National Security Archive Electronic Briefing Book No. 449." Posted 11 December 2013, https://nsarchive2.gwu.edu/NSAEBB?NSAEBB449/, viewed 10 March 2021.

39 https://fas.org/irp/offdocs/wmd_report.pdf,   The   Robb-Silberman Commission Report.

40 Operations Northern and Southern Watch. https://en.wikipedia.org/wiki/Operation_Southern_Watch.
https://en.wikipedia.org/wiki/Operation_Northern_Watch, viewed 27 June 2021.

41 www.cnn.com/2013/10/30/world/meast/iraq-weapons-inspections-fast-facts/index.html.
42 Yet, in spite of all the planning, everything did not go perfectly with Y2K. https://fas.org/sgp/news/2000/01/nyt010200.html, viewed 25 June 2021.
43 https://crisp.nus.edu.sg/~research/tutorial/dmsp.htm.
44 James Risen and Tim Weiner. "Indian Nuclear Tests," *NYT*, 25 May 1990.
45 Jeffrey Richelson, *Spying on the Bomb* (New York, NY: Norton, 2006), pp. 432 and 444–445. Richelson's Chapter 11, Pokhran Surprise, outlines the history of how India learned how US intelligence, particularly its imagery intelligence, monitored its nuclear weapons program, and changed its test preparations accordingly before the 1998 test.
46 Bissell, Jr. Richard M. *Reflections of a Cold Warrior: From Yalta to the Bay of Pigs,* With Jonathan E. Lewis and Frances T. Pudlo (New Haven, CT: Yale University Press, 1996), pp.105. cited in Chapter Five, footnote 14, James Reber, interview by Jonathan Lewis, tape recording, Chevy Chase, MD. 11 February 1994; O'Connor. *NPIC: Seeing the Secrets and Growing the Leaders*, pp. 105–107.
47 Most analytic work in the intelligence community can be considered a level-of-effort program.

BIBLIOGRAPHY

ACIU Map showing azimuths of German V-1 launch sites in August 1944. (The Medmenham Collection).

Babington-Smith, Constance. *Evidence in Camera: The Story of Photographic Intelligence in the Second World War* (Phoenix Mill, Gloucestershire: Sutton Publishing, 2004).

Bissell, Jr. Richard M. *Reflections of a Cold Warrior: From Yalta to the Bay of Pigs,* with Jonathan E. Lewis and Frances T. Pudlo (New Haven, CT: Yale University Press, 1996), pp. 105, cited in Chapter Five, footnote 14, James Reber, interview by Jonathan Lewis, tape recording, Chevy Chase, MD. 11 February 1994.

Brugioni, Dino A. *Eyeball to Eyeball: The Inside Story of the Cuban Missile Crisis* (New York, NY: Random House, 1990).

———*Eyes in the Sky: Eisenhower, the CIA, and Cold War Aerial Espionage* (Annapolis, MD: Naval Institute Press, 2010).

Burtch, R. Class Notes Sure 340, 2009, *History of Photogrammetry*, Ferris State University.

CIA. (Ad Hoc Requirements Committee on Project AQUATONE (ARC) Minutes of Meetings Held in Room 121 Administration Building, Central Intelligence Agency, at 1200, 1 June 1956 and at 1400, 4 June 1956. SAFC 678 Approved for Release 2000/08/21/ RIA-RDP33-02415A000100100056-7.

———U.S.S.R.: *Nuclear Accident Near Kysthym in 1957–58*, An Intelligence Assessment. National Foreign Assessment Center, SECRET, SW81-10102, October 1981. CIA Historical Review Program Release as Sanitized 1999.

www.cia.gov/readingroom/.

www.cnn.com/2013/10/30/world/meast/iraq-weapons-inspections-fast-facts/index.html.

Conyers-Nesbitt, Roy. *Eyes of the RAF: A History of Photo-Reconnaissance* (Phoenix Mill, UK: Sutton Publications, 1996).

Cottrell, C. M. and Milton Glicken. "Biography of a C-4 Stereoplanigraph." www.mdshs.org/Biography%20of%20a%20C-4.html.

https://crisp.nus.edu.sg/~research/tutorial/dmsp.htm.

Dobbs, Michael. "Tracing the Nuclear Warheads, One Minute to Midnight: Kennedy, Khrushchev, and Castro on the Brink of Nuclear War." Posted 18 June 2008. https://nsarchive2.gwu.edu/nsa/cuba_mis_cri/dobbs/warheads.htm.

Eder, J.M. *History of Photography* (New York, NY: Colombia UP, 1945).

https://fas.org/sgp/news/2000/01/nyt010200.html.

Finnegan, Col. Terrence, J. *Shooting the Front: Allied Aerial Reconnaissance and Photographic Interpretation in the Western Front – World War I* (Washington, D.C.: NDIC Press, 2006).

Hardesty, Von. *Camera Aloft: Edward Steichen in the Great War* (Cambridge, UK: Cambridge UP, 2015).

https://history.state.gov/milestones/1969-1976/salt.

Jones, R. V. *Most Secret War* (London, UK: Hamish Hamilton, 1978).

Kaye, Thom. "Floyd W. Hough," http://disturbedgeographer.com,

www.mdshs.org/Biography%20of%20a%20C-4.html.

Medvedev, Zhores. *Nuclear Disaster in the Urals* (New York, NY: Norton, 1979).

Miller, Greg. "Behind the Lines: The Untold Story of the Secret U.S. Mission to Capture Priceless Mapping Data Held by the Nazis," *Smithsonian Magazine*, November 2019, pp. 64–78.

Naftali, Timothy and Aleksandr Fursenko. *One Hell of a Gamble: Khrushchev, Castro, and Kennedy, 1958–1964* (New York, NY: W.W. Norton & Co. 1998).

Newhouse, John. *Cold Dawn: The Definitive History of SALT* (New York, NY: Reinhart and Winston, 1973).

O'Connor, Jack. *NPIC: Seeing the Secrets, Growing the Leaders* (Alexandria, VA: Acumensa Press, 2015).

Pooley, Ruth. "Wild War Stories," *The Medmenham Magazine*, Spring 2021, pp. 26–27.

———"Wild War II," *The Medmenham Magazine*, Autumn 2021, pp. 38–41.

Powys-Lybbe, Ursula. *The Eye of Intelligence* (London: William Kimber, 1983), p. 61.

Richelson, Jeffrey T. *Spying on the Bomb* (New York, NY: Norton, 2006).

Risen, James and Tim Weiner. "Indian Nuclear Tests," *NYT*, 25 May 1990.

Svetlana Savranskaya and Thomas Blanton (ed.) with Anna Melyakova. "Last Nuclear Weapons Left Cuba in December 1962. National Security Archive Electronic Briefing Book No. 449." Posted 11 December 2013, https://nsarchive2.gwu.edu/NSA EBB?NSAEBB449/.

Stanley, Col. Roy M. *World War II Photo Intelligence* (New York, NY: Scribners, 1981).

Stanley, Col. Roy M. *V Weapons Hunt: Defeating German Secret Weapons* (Barnsley, UK: Pen and Sword Military, 2010).

Stewart, Paul. *Medmanham: Anglo-American Photographic Intelligence in the Second World War, Volume 1* (submitted for the Doctor of Philosophy at the University of Northampton, 2019).

https://en.wikipedia.org/wiki/Operation_Northern_Watch.

https://en.wikipedia.org/wiki/Operation_Southern_Watch.

Williams, Allen. *Operation Crossbow: The Untold Story of the Search for Hitler's Secret Weapons* (London, UK: Random House, 2013).

CHAPTER 6

Tracking

INTRODUCTION

Tracking, the fifth and final intellectual activity of geospatial intelligence, deals with the human propensity to move. It began during World War I with the sporadic tracking of large numbers of men and large amounts of equipment. The World War II attempts to track individual mobile targets from aircraft began with the British and Allied efforts to locate the V1 and V2 German missiles. While the British-led Allied Central Interpretation Unit made progress at locating a number of fixed launchers, it remained unsuccessful at understanding the movements and stopping the effective use of these weapons, much less following them in real time.

In the early 1980s, partially in response to US successes at discovering, identifying, and comprehending its fixed missile infrastructure, the Soviet Union deployed the SS-20 mobile missiles, but satellite collection against this missile system was no more effective than World War II aircraft collection had been against the V-1 or V-2. On any given day, neither aircraft nor satellites could image a large enough area nor collect over a long enough period of time to track these missiles. The US had located the SS-20 infrastructure and bases, identified its order of battle, and recognized some of its camouflage practices, but efforts to track this missile force in real time fared no better than World War II British efforts against the V-2. The challenges of tracking missile systems persisted through the next war, when the Allied forces tried unsuccessfully to find and destroy SCUD missile launchers in Iraq during Operations Desert Storm and Desert Shield.

The first successful tracking with geospatial technology took place in Bosnia in the 1990s. The Central Intelligence Agency (CIA) flew an experimental unmanned imaging aircraft—the GNAT-50, called a drone on account of its whining engine. This remotely piloted aircraft introduced the possibility of persistent surveillance with high resolution digital video images, a capability to loiter over an area for many hours, and its relative invisibility. Its suitability for following military forces in mountainous terrain and individual pieces of equipment for hours at a time introduced the modern age of tracking.

CIA's success over Bosnia in tracking individual pieces of equipment with drones spawned an experiment in using this technology to track terrorists in Afghanistan. Initially a larger and better equipped follow-up drone, the Predator, was unarmed, but after observing an individual believed to be Osama Bin Laden in the summer of 2000, and the subsequent terrorist bombing of the USS Cole in Yemen in October 2000, the US decided to arm the Predator drone.

After Al Qaeda attacked in the US on 9/11, CIA started to use drones for both intelligence gathering and operational strikes in the pursuit of terrorists and ultimately Osama bin Laden. Drone collection became the underpinning for "activity-based analysis," a form of geospatial intelligence that relied on persistent surveillance, GIS-based reporting, and the rapid acquisition of spatially registered cell phone intercepts. Activity-based intelligence requires large expenditures of human attention and data storage, particularly when it is used against individuals as the activity has to be discovered and comprehended before it can be tracked. The combination of this tracking technology and analytic methodology would eventually enable the finding, monitoring, and attack raid at Osama Bin Laden's Abbottabad compound in 2015.

The growth in geospatial technology and the reliance of global social media on location-based technology enabled precise tracking by anyone with a computer terminal. While individuals cannot track in real time, as governments do, they still can accomplish it rapidly enough to defeat governmental attempts to deny geospatial information, deceive with false reporting, or create false narratives about governmental activities. In 2014, Bellingcat, a private media organization composed of a few tenacious and talented individuals, used geospatial analysis and technology to track and locate the Russian officers and military equipment that shot down Malaysian Airlines Flight 17, a commercial airliner, over Ukrainian airspace.

TRACKING

Tracking is the last intellectual activity in geospatial intelligence after envisioning, discovering, comprehending, and recording. In spite of the diminishing costs of geospatial technology, tracking remains the most expensive activity with regard to human attention, duration of persistent coverage, data storage, and round-the-clock recording requirements.

Most imagery collection efforts, from World War I through the end of the film-return era in 1985, were from hours to months old when first looked at by an interpreter or analyst. Consequently, they captured only a moment in time. Subsequent analysis of a captured moment in time sufficed in an era of static warfare, unit movements reliant on shipboard, railroad, or motorized schedules, or construction of static facilities that took months or years to complete.

The launch and orbiting of the KEYHOLE-11 (KH-11), the first satellite to provide daily digital imagery for intelligence, beginning in January 1977, greatly increased the volume of imagery. This imaging satellite was part of a system that included relay satellites to accelerate the return of the imaging data from space,[1] but the orbiting time and the processing time meant that there would always be a lag between the time of imaging and the earliest time when an analyst could look at the image on a computer screen or light table. Another lag would follow until the intelligence from the imagery could reach those who needed it to act quickly or think differently in response to it.

In sampling terminology, as long as the periodicity of the observation was within the schedule of the construction or activity, geospatial intelligence could provide warning. This model worked reasonably effectively from World War I through the first three decades (1950s–1970s) of the Cold War. However, the Soviet invasion of Czechoslovakia in 1968 and the 1973 Arab–Israeli war were notable examples of the failure of film-return systems to enable imagery analysts to warn of rapidly developing events. At the end of the 1970s and in the early 1980s the Soviet Union began to deploy a missile system that forced US imagery analysts to change how they thought and worked.

The Soviet SS-20, a truck-mounted, solid-fueled missile armed with three nuclear warheads, challenged the intelligence community.[2] The SS-20 made the US Intelligence Community recognize that it needed to accelerate its operations. Because this missile was solid-fueled, it could be moved to a location, leveled on its launcher jacks, employ its communications and theodolites, and within 30 minutes be capable of launching three nuclear warheads into Western Europe. In response to its deployment, the Reagan administration deployed US Pershing solid-fueled intermediate-range ballistic missiles into Western Europe. While this deployment aimed the same kind of weapon that could be deployed and launched on a similar schedule from any European location into the Soviet Union, the mutual deployment left the US and the USSR with the same geospatial intelligence challenge: how to track relocatable strategic weapons in a timely manner to reduce or prevent their effectiveness?

The US had been successful at analyzing Soviet missile bases and comprehending their organization and the missile system infrastructure. For the new mobile missiles, it used the same analytic techniques. From the unique garages in which the Soviet Union stored its SS-20s, the US Intelligence Community could identify the bases and the number of deployed missile launchers. As the missile launchers left their bases to deploy in the field, the analytic challenge became more and more complex. In the field, the launchers used camouflage, terrain features, and movement to avoid detection and observation by US satellites.

Mobility may have been the most effective deterrent, because to hunt for an SS-20 required looking through a large amount of satellite search imagery. As any imaging satellite can only look at a single point on the earth's surface for a maximum of eight minutes on any one orbit, there were multiple hours every day during which the SS-20 crews could relocate, redeploy their communications and security, and camouflage themselves. Thus, even with near-real-time imagery, the analysts were searching for mobile missiles on imagery that would be hours old when they received it. A small number of missile launchers could deploy outside an SS-20 base, and, with prior knowledge of the US satellite orbits,[3] move during the intervals between satellite overflights to different locations and re-camouflage themselves before the next opportunity for a US satellite to image their operating area. Imagery analysts had to search large areas of the USSR rapidly in hopes of detecting these weapons.

While the Soviet response to this question of how to locate US mobile missile systems remains classified or undocumented, the US began a number of intelligence experiments in the 1980s. For the first time, it connected its imagery tasking system to other types of intelligence collection,[4] so an imagery observation could initiate taskings for other kinds of US intelligence collection that were geospatially registered. The search for SS-20s remained a US intelligence priority until 1986 when one office in the intelligence community searched 60% of the USSR to establish a baseline for the SS-20 force before the Reykjavik summit where President Reagan met with Premier Gorbachev. After the October 1986 Reykjavik summit, the US and the USSR entered into the Intermediate-Range Nuclear Forces (INF) Treaty negotiations, and the US based its negotiating position on the results of that imagery search.[5]

Unknown at that time in the US, the Reagan administration's announcement of the Strategic Defense Initiative (SDI) forced the civilian leadership in the Soviet Union to recognize that it couldn't compete technically or economically in developing a military weapons system that could neutralize the SDI.[6] Gorbachev recognized the economic costs, and his arms control initiatives provided evidence of his recognition. The INF treaty, an outcome of the Reagan–Gorbachev summit at Reykjavik, began to shrink the Cold War tracking challenge of mobile missiles.[7]

THE SCUD HUNT

Although the 1987 INF treaty ended the SS-20 tracking challenge for the US, four years later a similar challenge developed in another country with older Soviet mobile missile technology. After Saddam Hussein invaded Kuwait in August 1990, the US organized a global military response to expel the Iraqi military forces. The UN military coalition's campaign, popularly known as Desert Shield/Desert Storm, deployed forces from August 1990 until February 1991. During the Desert Shield part of the operation from 17 January until 28 February 1991,[8] the Iraqi weapon against which the US forces had the least effective defense was the SCUD, a Soviet mobile tactical ballistic missile with a range of less than 500 miles.

Initially the Soviet Union deployed the SCUD in 1957. Its design had been derived from the captured German V-2 technology that had challenged the British in World War II. After its initial deployment the SCUD had been improved a number of times to increase its range. Iraq had begun receiving SCUD missiles from the USSR in 1974,[9] and it had gained experience operating and launching the SCUDs during the 1980–1988 Iran–Iraq war. The Iranian military also had SCUDs and used them against Iraq in that war. During Desert Storm and Desert Shield, Iraq successfully launched 46 SCUD missiles into Saudi Arabia and 42 into Israel. The coalition forces, predominantly the US and the British, searched for SCUD launch sites in Iraq with special forces on the ground and fighter aircraft flying day and night operations in Western Iraq. But the combination of the high speed of fighter aircraft and the limitations of satellite access meant that the Allied search for SCUDs in Iraq was no more effective than the World War II British search with reconnaissance aircraft for German V-1 and V-2 launch sites. The mobility of the Iraqi SCUDs defeated the UN coalition forces' attempt to locate and defeat this weapon.[10]

BOSNIA AND GLOBAL WAR AGAINST TERRORISM

After the collapse of the Soviet Union in 1991,[11] the governments of several of its former European satellite states also collapsed. In two of these states, Yugoslavia and Albania, when social order collapsed, ethnic conflict ignited. In the former Yugoslavia, which collapsed in 1991–1992,[12] tracking the combatants challenged the geospatial analysts. All the military forces of each ethnic group—Bosnians, Croatians, Albanians, and Serbs—used identical former Soviet military equipment. In many areas, different ethnic groups with religious differences and a history of conflict extending for more than half a millennium were intermixed in the mountainous patchwork of small villages.

The US introduced a new technology that changed the Intelligence Community's approach toward tracking. In a technological experiment, CIA began to monitor the conflict with a pilotless aircraft, or drone, equipped with a video camera, a direct downlink, and a small engine that propelled this drone at 90 miles per hour. Initially, the experiment had technical challenges with terrain masking the camera feed and the control link.[13]

The Air Force joined CIA in flying subsequent drone missions over Bosnia and improved the drone technology, particularly its command, control, and communications. The surveillance drone flew at about the same speed as World War I observation aircraft, and could loiter for hours over the same area. It was slow enough, small enough, and quiet enough to be very difficult for air defense radars to detect it electronically or ground forces to detect it visually. It also carried infrared sensors so that it could image at night. Its technology and attributes enabled it to be very effective at tracking ground forces for hours on end.

This new capability meant that in Bosnia, the Allied geospatial intelligence effort could accomplish much more than previously. The Predator augmented near-real-time satellite intelligence with real-time video surveillance that was relayed by satellites across the globe to a classified location. It could track unit movements over multiple hours instead of for the few minutes that satellites are restricted to. Also, if a military unit remained in a certain area, the drone could loiter and observe the activities of the unit.

Most importantly, if the drone observed atypical or suspicious activity, it could follow individual vehicles or groups of people. The combination of increased resolution and duration enabled drone reconnaissance and surveillance aircraft to identify individual vehicles and track them for multiple hours. The US State Department would later take the intelligence obtained from this capability to identify military activity against civilians. Later at the World Court the activity captured with this sensor would be proven to be a war crime related to ethnic cleansing.

On a positive note, the ability to watch moving military forces continuously in the Balkans conflict helped US diplomatic negotiators throughout the Dayton negotiations to persuade all parties in the negotiations of the US ability to monitor troop movements and peaceful civilian migrations. This technology also helped the negotiators manage expectations for both sides. Neither side could move forces without the US observing and tracking the changes. As the US began to focus on individual Bosnian Serb units and their leaders who had been orchestrating the systematic destruction of Kosovar Muslim villages and in a few places the systematic extermination of people, the drones were used in the tracking and capture of some of these war criminals.

After the Balkans War, the US continued to improve this new surveillance capability. After the initial CIA experiment with the GNAT 750,[14] the US Air Force developed the Predator which carried much greater communications capability and better sensors. The intelligence value of the drone grew proportionally as the US became more aware of the terrorist threat of Al Qaeda, and its leader Osama bin Laden.

Following the African Embassy bombings in 1998, CIA had added analytic resources on Osama bin Laden and Al Qaeda. By the summer of 2000, the Air Force, CIA, and the National Security Council designed a "Summer Project," to fly an unarmed Predator drone over Afghanistan from a former Soviet airbase in Uzbekistan. The aim of the "Summer Project" was to attempt to locate Osama bin Laden in Afghanistan. For this experiment the US had created an experimental global architecture with nodes in Uzbekistan, Germany, and the US. This architecture enabled the US to transmit Predator video feeds nearly instantaneously to each of the nodes.[15]

The slow-flying Predator could take multiple hours to arrive at an area of interest and loiter over that area from 20 to 24 hours. In the "Summer Exercises," this loitering capability introduced the possibility of persistent surveillance. In conjunction with other digital intelligence sources, persistent surveillance enabled geospatial analysts to establish a pattern of life over time at a particular location. Given such a pattern of life with documented times of routine activity—arrivals, movements, and departures—the patterns of communication could be tied to the movements of individuals, and maps of the human networks under observation could be created. This form of analysis, called network analysis or activity-based analysis, developed rapidly.[16]

On 27 September 2000, during the Summer Exercises, at Tarnak Farms in Afghanistan, the Predator sensor and those watching its feed on video monitors observed a tall man in white robes surrounded by other men paying deference and providing security. Although at the time they had no specific intelligence placing him at this location, the geospatial analysts believed that the man might have been Osama bin Laden. A short-lived debate developed in the US Intelligence Community about what to do with this unconfirmed evidence of bin Laden's current location.[17] While the "Summer Exercises" were ongoing, planning efforts had been underway to arm the Predator with modified Hellfire missiles. In September 2000, the US government decided not to take any action at Tarnak Farms, but two weeks later, after suicide bombers in Yemen attacked the USS Cole on 12 October the debate ended. After the USS Cole bombing, the US Air Force accelerated its program to arm the Predator with air-to-surface missiles.

In the year before 9/11, the US Air Force increased the size of its Predator drone force and began to arm some of them. CIA's Counterterrorism Center, which had been among the advocates for military action at Tarnak Farms in 2000, lobbied to have armed Predators flown over Afghanistan. And during this year, Al Qaeda was also planning its 9/11 attack on US targets.

After 9/11, the US turned its armed and unarmed drone technology on individuals and groups in Afghanistan supporting the Taliban and

Al Qaeda. In 2003 it brought drone technology to Iraq. Initially the US flew drones in Afghanistan from late 2001 through 2002 in Operation Enduring Freedom. The drones provided persistent surveillance and assisted in target identification for laser-guided weapons. The activity-based analysis that grew out of drone technology began to change the nature of drone missions.

Drones provided the luxury of long observations of activity for intelligence, but for many geospatial analysts looking at full-motion-video often felt like the misery of boredom. In the global war on terrorism, the US developed its capability to identify an individual's digital signature through a cell phone or computer. And if that individual moved from one location to another, the US could follow his or her "digital exhaust." But when the individual arrived at the second location, initially analysts would not know if he or she were there for a family visit, for a meal, for a needed rest, a meeting, a religious holiday, or an unknown reason. Drone missions began to resemble aerial versions of police stakeouts with the observers a continent away from the observed. As all this geo-tagged video imagery sent back to the US had to be recorded for potential future retrieval if more intelligence was obtained from different sources, These drone missions were expensive in human attention and digital storage. Many times, these missions had no clearly identifiable outcome, but in some cases, after other intelligence sources were obtained, they did.

In the US counterterrorism effort, the US national intelligence priority at the time, the intelligence and geospatial community was willing and able to pay the resource and attention costs for these missions. But the resource costs of this approach were affordable only for the highest priority missions. In 2008, the final steps in the hunt for bin Laden illustrated the willingness of the US to pay the cost of tracking.

After the discovery of the Abbottabad compound, and the judgment of its potential association with Osama bin Laden, the surveillance became persistent to establish the number of inhabitants in the compound and their pattern of life. All applicable US geospatial intelligence collection resources—drones, commercial satellites, and government satellites—were incorporated into this collection plan. The Intelligence Community obtained sufficient detail about the Abbottabad compound for the National Geospatial-Intelligence Agency to build a scale model used first in National Security Council meetings to discuss and approve the raid and later in US Special Operations Command to plan for the Abbottabad raid (Figure 6.1).

In spite of all the geospatial sources and all other intelligence sources commanded by the highest US strategic national priority, until the culmination of the raid, just as at Tarnak Farms in 2000, uncertainty remained about whether one of the men at the Abbottabad compound was bin

FIGURE 6.1   Photograph of scale model of Usama bin Laden's compound in Abbottabad, Pakistan.

Source: Courtesy of the National Geospatial-Intelligence Agency Historic Research Center (Abbottabad Compound Model, NGA 2011.25.01).

Laden. The Abbottabad raid remains the most detailed example of a government intelligence organization tracking an individual with geospatial and other sources of intelligence. It also provides a clear example of the uncertainty that surrounds an intelligence call even when the issue had commanded all available collection resources and analytic attention.

FORENSIC TRACKING

While live tracking in the present may remain a wholly governmental activity, restricted to the highest priority targets, in 2014, a private organization conducted a geospatial search to identify mass murderers that resulted in a World Court indictment.[18] On 27 July 2014, a Russian air-defense unit deployed on Ukrainian territory used its radar to track a civilian airliner over Ukraine. The Russian air-defense unit launched a surface-to-air missile and shot down Malaysian Airways Flight 17, a regularly scheduled flight from Amsterdam to Kuala Lampur. All 298 passengers and crew aboard were killed.

The Russian government denied any involvement, but a private journalistic organization, Bellingcat, had a network of volunteer social media experts and GIS experts, as well as leadership willing to purchase commercial imagery. Very quickly, it set in place an unclassified collection plan to discover what actually happened over the Ukraine. Bellingcat envisioned that they could create a map of the events of 27 July 2014 and chronicle them moment-by-moment with geospatially registered data. The significant difference with this retrospective tracking is that Bellingcat accomplished it without any government resources or intervention, and used multiple sources of volunteered geospatial information.

Bellingcat's methodology was to use cell phone images and videos to capture a large mosaic image of the area of the shootdown and the crash. By using geospatially registered metadata from these cell phone images and videos, the Bellingcat analysts were able to plot all the still and video images, ascertain their field of view and orientation from the locations of the cellular phones as well as to record the time of each image.

From this initial volunteered collection, Bellingcat analysts chronicled a different sequence of events than what either the Ukrainian government or the Russian military described. As the amateur Bellingcat analysts worked, they plotted the Russian surface-to-air missile trajectory from its launch south of Snizhne to its point of impact over the southern Ukraine. This aerial track was also geo-registered and time-stamped, so it could be correlated with the terrestrial images from the cell phones, unclassified flight data, and the recovered flight recorders.

After the missile launch location had been established, the next stage of analysis was the traditional imagery analysis process of negation, or searching through all the images near the location of the launch, as well as going back in time to look for images that established when the Russian Air Defense Unit went into the Ukraine and deployed to that specific location. In the process of negating this unit's location, a Bellingcat analyst who had served in an air defense unit had analyzed enough images of Russian air defense equipment to be able to distinguish among individual transporter-erector-launchers (TELs) the specific piece of equipment that carried the Russian surface-to-air missile that had shot down MH17.

The analyst was able to identify and locate each TEL in the area of the missile launch. He also established that one image showed the specific TEL leaving the launch area shortly after the time of the launch with one of its three missiles missing. Finally, the analyst had studied the images of all the TELs in that Air Defense Unit so well that he could distinguish individual characteristics among the vehicles. By this point in the investigation, his knowledge of the unit's equipment was so comprehensive that he could identify this individual TEL, even though the Russians had repainted all the TELs in this unit so as to make the identification more difficult.[19]

The Bellingcat analysis of the MH-17 shootdown illustrates the ability of nongovernmental organizations, using crowdsourcing and volunteered geographical information, to perform sophisticated geospatial analysis with only a modest expenditure for collection, processing, and disseminating the results of their analysis. The work and organization of Bellingcat illustrates the democratization of geospatial forensic tracking, and it provides a pathway for more of this work. Independently of its organizational location, geospatial intelligence can increase the potential that individuals and organizations can be held accountable for actions. Yet the movement of geospatial intelligence from being the work of nations to being the work of individuals or small groups raises future issues that deserve consideration.

NOTES

1 *The Era before KENNAN/KH-11, Section 2 KENNAN System.* pp. 84, 94, 124 (Chantilly, VA. Center for the Study of National Reconnaissance, 2021). Approved for Release 2101/07/09 C0509763.

2 O'Connor. *NPIC.* pp. 203–218.

3 William Kampiles, a CIA employee, had sold a copy of the KH-11 technical manual to a Soviet intelligence officer in 1978. www.courtlistener.com/opinion/371921/united-states-v-william-peter-kampiles/, viewed 17 December 2021.

4 O'Connor. *NPIC.* p. 214.

5 O'Connor. *NPIC.* pp. 203–218.

6 Hoffman, David E. *The Dead Hand: The Untold Story of the Cold War Arms Race and Its Dangerous Legacy* (New York, NY: Doubleday, 2009), p. 280.

7 https://2009-2017.state.gov/t/avc/trty/102360.htm.

8 https://history.state.gov/milestones/1989-1992/gulf-war, viewed 15 November 2021.

9 https://fas.org/nuke/guide/iraq/missile/scud.htm, viewed 25 June 2021.

10 Gordon, Michael R. and General Bernard E. Trainor. *The General's War: The Inside Story of the Conflict in the Gulf* (Boston, MA: Little Brown, 1995), Chapter 11, pp. 227–248, remains the best analysis of the intelligence and operational difficulty in tracking mobile weapons systems in a wartime environment.

11 https://history.state.gov/milestones/1989-1992/collapse-soviet-union, viewed 15 November 2021.

12 https://history.state.gov/milestones/1989-1992/breakup-yugoslavia, viewed 15 November 2021.

13 Boyle, Michael, J. *The Drone Age: How Drone Technology Will Change War and Peace* (New York, NY: Oxford University Press, 2020), pp. 50–53.

14 Whittle, Richard. *Predator: The Secret Origins of the Drone Revolution* (New York, NY: Henry Holt and Co., 2014), pp. 68–73, 81–82.

15 Whittle, *Predator*, pp. 143–163.

16 The book that describes this methodology best is Patrick Biltgen and Steven Ryan's, *Activity-Based Intelligence: Principles and Applications* (Boston, MA: Artech House, 2015).

17 Bierbauer, Alec and Col. Mark Cooter, USAF (ret) *Never Mind, We'll Do It Ourselves. The Inside Story of How a Team of Renegades Broke Rules, Shattered Barriers, and Launched a Drone Warfare Revolution* (New York, NY: Skyhorse Publishing, 2021). The participant's narrative of this event is captured on pages 114–118.

18 Higgins, Eliot. *We Are Belling Cat: Global Crime, Online Sleuths, and the Bold Future of News.* (New York, NY: Bloomsbury, 2021), Chapter 2: "Becoming Belling Cat: A Team of Detectives Takes Shape," pp. 63–109.

19 Higgins. *We Are Belling Cat*, pp. 72–88, 94, 97–102.

BIBLIOGRAPHY

Bierbauer, Alec, and Col Mark Cooter with Michael Marks. *Never Mind, We'll Do It Ourselves: The Inside Story of How a Team of Renegades Broke Rules, Shattered Barriers, and Launched a Drone Warfare Revolution* (New York, NY: Skyhorse Publishing, 2021).

Biltgen, Patrick and Steven Ryan, *Activity-Based Intelligence: Principles and Applications.* (Boston, MA: Artech House, 2015).

Boyle, Michael, J. *The Drone Age: How Drone Technology Will Change War and Peace* (New York, NY: Oxford, 2020).

www.courtlistener.com/opinion/371921/united-states-v-william-peter-kampiles/.

www.fas.org/nuke/guide/iraq/missile/scud.htm.

Frantzman, Seth J. *The Drone Wars: Pioneers, Killing Machines, Artificial Intelligence, and the Battle for the Future* (New York, NY: Post Hill Press, 2021).

Gordon, Michael R and General Bernard E. Trainor. *The General's War: The Inside Story of the Conflict in the Gulf* (Boston, MA: Little Brown, 1995).

Higgins, Eliot. *We Are Bellingcat: Global Crime, Online Sleuths, and the Future of News* (New York, NY: Bloomsbury, 2021).

https://history.state.gov/milestones/1989-1992/gulf-war.

https://history.state.gov/milestones/1989-1992/collapse-soviet-union.

https://history.state.gov/milestones/1989-1992/breakup-yugoslavia.

https://2009-2017.state.gov/t/avc/trty/102360.htm Intermediate Range Nuclear Forces Treaty.

Hoffman, David E. The Dead Hand: *The Untold Story of the Cold War Arms Race and its Dangerous Legacy* (New York, NY: Doubleday, 2009).

NGA. Photograph of scale model of Usama bin Laden's compound in Abbottabad, Pakistan (Courtesy of the National Geospatial-Intelligence Agency Historic Research Center (Abbottabad Compound Model, NGA 2011.25.01).

NRO. Center for the Study of National Reconnaissance. *The Era before Kennan/KH-11.* www.nro.gov/Portals/65/documents/foia/declass/HISTORICALLY%20SIGNIFICANT%20DOCs/NRO%2060th%20Anniversary%20Docs/SC-2021-00002_C05097836.pdf.

O'Connor, Jack. *NPIC: Seeing the Secrets, Growing the Leaders* (Alexandria, VA: Acumensa Press, 2015).

Whittle, Richard. *Predator: The Secret Origins of the Drone Revolution* (New York, NY: Henry Holt and Co., 2014).

CHAPTER 7

Contemporary Geospatial Intelligence

INTRODUCTION

Two complementary and simultaneous efforts—technical progress and human ingenuity—shape and exemplify modern geospatial intelligence. They relate to an earlier attempt in the history of ideas to reconcile two opposing perspectives on the future. More than one hundred years ago, Henry Adams attempted to use a technology and a religious figure, "The Dynamo and the Virgin," in his Pulitzer-prize winning *The Education of Henry Adams*,[1] to express the distinction between the human and the technical.

The current state of geospatial intelligence can be described by referring to the contexts of faith and science that Adams used. Among geospatial intelligence practitioners, many believe in the historical evolution of the profession. Their approach is linked to the tradition of imagery analysis and photointerpretation and reflects a faith in human intellect and imagination as the key to solving intelligence problems. This view of geospatial intelligence believes that incrementally more sources with better resolution, greater swath width, and increasing periodicity of collection will eventually get to a state that has been described as "the panopticon," "the all-seeing eye," and ubiquitous coverage. But this state usually is described as being achieved over a relatively small geographical area—a pair of countries, a region, or a part of a single continent. The other perspective on geospatial intelligence in the early part of the second decade of the 21st century has its basis, like Adams's dynamo, in scientific progress.

Some who practice geospatial intelligence rely on and experiment with using parts of Maxwell's electromagnetic spectrum to provide information about events in the world. These advocates use incremental

DOI: 10.4324/9781003436836-8

engineering to increase coverage of the planet by sensors of every type. More and more of the planet is collected daily with more sensors and a wider variety of sensors so that a greater variety of the spectrum is being used to sense and generate geospatial data more frequently. The increasing amounts of all kinds of data are being imaginatively used to inform geospatial analysis. The increasing amount of data, like Adams's accelerating dynamo, contributes to the general global good.

The tension in contemporary geospatial intelligence lies between what human analysts ought to do for geospatial intelligence and what geospatial technology ought to do for human analysts. This tension is not new in geospatial intelligence nor in American cultural history. Adams contrasted the power of electrical technology symbolized by the Dynamo, and his symbol of the power of imaginative envisioning was the Virgin, the religious figure of Mary, mother of Christ, as depicted in the western tradition of Christian art shown in the Hall of Art at the World's Fair. A hundred years later, those engaged in geospatial intelligence currently are trying to comprehend an important relation: how do the uniquely human powers to envision, discover, record, comprehend, and track human activities, which Adams associated with art and the cultural tradition, relate to and combine with the complementary digital electronic power, which Adams would associate with science to improve the human condition?

Having outlined the five intellectual habits of geospatial intelligence and the technologies that enable them, this chapter analyzes the application of geospatial intelligence to two contemporary intelligence issues: the war in Ukraine that began in 2022 and the impact of climate changes on the planet. The first is a traditional military conflict between two nations, and it extends the more than one-hundred-year tradition of geospatial intelligence as a foundation of military intelligence. The second is uniquely modern, and it involves trying to define a global threat and provide geospatial intelligence in order to inform future policies and actions.

The intelligence issues surrounding climate change are unprecedented, but well suited to geospatial intelligence. The problem is global with no agreed-upon indicators of future events, nor is there any international or scientific agreement about what areas of the earth should be monitored or at what frequency.

UKRAINE

Many Russian military actions in the Ukraine war, such as the siege of the Avrostal Steel Complex, resemble World War I combat and others resemble post-DESERT STORM "precision" warfare. The extended combat in the Donbas and Luhansk areas of eastern Ukraine, continuing from 2014 through 2022, reflects the positional, entrenched mode of ground combat that has existed since 1914. In other parts of Ukraine, the war, at least as practiced by the Ukrainian military, is a digitally precise modern war.

The geospatial envisioning that the Ukrainians have brought to prominence is a new use for commercial electro-optical satellite imagery. For the first time, one nation at war has used available commercial imagery from US and other countries to undercut the information operations of another nation. Since late 2021, Ukraine has analyzed on a regular basis western commercial imagery to discover and forecast new Russian military activity. Ukraine employs geospatial intelligence to forecast and warn not only its military but also the global media audience about changes in Russian military dispositions in Russia, Belarus, and Ukraine. It has also used analysis based on this imagery to undercut the Russian disinformation campaign that began in 2021 prior to their invasion of Ukraine in February 2022 and the disinformation campaign about the civilian killings in Bucha.

In central and western Ukraine, near Kiev and Kharkov, both countries have fought the war with 21st-century precision weapons. Aircraft and air defense missiles of both nations have incorporated targeting technologies informed by the global positioning system (GPS) and geographic information systems (GIS). The Russian destruction of the Antonov 225 in February 2022 may have been their most prominent success. While 21st-century precision technology has enabled Ukraine to offset the numerical superiority of the Russian military, it has not enabled Ukraine to overcome the Russian numerical and equipment advantages.

Some signs of conflict, entrenched infantry positions and camouflaged artillery duels and barrages in the Donbas and Luhansk region very much resemble the combat of World War I. But other attacks by both sides demonstrate the power of geospatial intelligence. Certain mobile targets, such as the Russian cruiser, Moskva, and the Ukrainian Antanov-225 transport aircraft, have been effectively tracked, subsequently targeted, and destroyed. Fixed infrastructure targets—railways, road bridges, fuel, and ammunition depots have been effectively and accurately targeted by both sides. Yet, as happened in World War I, there has been imprecise and indiscriminate shelling by both armies.

Both countries demonstrated geospatial comprehending in their use of precision targeting. During the Cold War the Soviet Union built much of what is now the Ukrainian military infrastructure—airfields, garrisons, fuel storage depots, and road and railroad networks, as well as conventional and nuclear power plants. Consequently, their comprehension of these targets has been developing for decades. Ukraine, in its ability to thwart numerically superior and better armed military forces around Kiev, demonstrated its comprehending of Russian forces and tactics and an effective defense based on tracking Russian force relocations.

As used by Ukraine, commercial geospatial intelligence has remained visible in world media through the conflict. Additionally, Ukraine is also using commercial geospatial intelligence for conventional military uses such as situational awareness, battle damage assessment, and analysis of the status of forces. The Ukrainian request for synthetic aperture radar (SAR) attests to this,[2] as Ukraine has not used commercial SAR in its information operations in the media. In bad weather and at night SAR, as it does not need sunlight to function, is very effective at tracking changes in force disposition. Also, both Ukraine and Russia continue to use geospatial intelligence obtained through drone operations for tactical situational awareness and tracking mobile targets.

In geospatial intelligence recording is the least visible intellectual activity. Yet its existence can be reasonably inferred by the targeting accuracy of some of both the Russian and Ukrainian weapons systems used by their ground, air, and naval forces. In particular, the Ukrainians have shown that they can effectively merge information from multiple geospatial sources—American and European commercially provided SAR and electro-optical sensors, as well as infrared and electro-optical sensors from Turkish drones. The Russians have their own recording processes, but they do not disclose their intelligence sources and methods.

The world media has highlighted the targeting capabilities of both countries. Both combatant nations have used video and still images of target strikes as evidence of their military effectiveness. As precision weapons depend on GPS systems for accuracy, geospatial information drives the use of these weapons. In the case of the Ukrainian attack and sinking of the Russian cruiser, *Moskva*, in the Black Sea, the Ukrainians have demonstrated their capability to track and strike mobile targets by using multiple reconnaissance and weapons systems in concert.

As both combatant nations attempt to control information about the progress of the war, the full extent of their use of commercial geospatial imagery likely will not be known for some time. Yet it is already possible to point out that the Ukrainians have frustrated the Russians in their attempt to use a traditional military practice. Russian, and previously Soviet, military doctrine advocated the use of "disinformation." The Ukrainians used electro-optical commercial imagery very effectively in late 2021 and early 2022 to delineate the location, size, and scale of the pre-invasion Russian military buildup in Belarus and western Russia. While the Russians tried to deny their advances in official statements and Russian media reports, the Ukrainian publication of commercial imagery that was disseminated to the global media, with daily updates, thwarted the disinformation campaign.

Commercial imagery defeated the Russian disinformation campaign by using historical coverage to make comparisons and by monitoring the staging areas and garrisons in Belarus and in western Russia daily to track the increasing arrivals of military equipment. When the Russian army tried to deny its killing of unarmed civilians in Bucha, a suburb of Kiev, the Ukrainians were able to use sequential collection of commercial imagery to demonstrate that the civilians had been killed many days prior and their corpses left in the streets for days before the Russians withdrew their forces from the area.[3] This use of commercial imagery to demonstrate the falsity of the Russian propaganda was also done in Chernobyl, Mariupol, Zaporizhzhia, and Kharkov.

At the time of this writing, in Spring 2023, the outcome of the war in the Ukraine remains unknown. But the Ukrainian use of commercial geospatial satellites has likely changed the nature of future ground combat elsewhere. The increased number of imaging satellites on orbit has greatly reduced the intervals between satellite collection windows. The gaps in satellite coverage are shrinking, and the early detection of military activity is increasing. Additionally, the size of the area observed by satellite imagery is growing.

One issue for geospatial intelligence will be the secondary effects of the war on the Ukrainian and global economy. The effects of the war on the regional agricultural sector, specifically the Ukrainian harvests in 2022 and 2023, will be assessed by multi-spectral sensors. Other significant Ukrainian economic sectors, such as mining, power generation, and fertilizer production, can also be monitored by satellites. The extent of battle damage to cities and other national infrastructure like highways, railroads, bridges, power plants, and dams can be measured from space. But the events of the war shown on commercial satellite imagery in the media are more dramatic and easier to grasp than economic or other non-military changes. These nonmilitary uses of geospatial intelligence, even though far less dramatic, will play a critical role in addressing the larger intelligence problem of global warming.

CLIMATE CHANGE OR GLOBAL WARMING

Geospatial intelligence is an important tool for identifying and measuring changes on the planet, and its importance is magnified when it can supply factual information to a critical and politicized problem. While everyone can perceive the effects of changes on the planet, the world seems unable to agree on what to call this problem, how to measure this problem, and what to do about this problem of man-made changes to the planet.

Evidence of climate change or global warming is everywhere, but that does not always help geospatial analysts know where or how to look next. Climate and environmental scientists are trying to relate all observable differences in different locations and climates, and one of their principal tools has been remote sensing. Since the launch of LANDSAT in 1972, more than 50 years ago, digital imaging satellites have been essential in measuring, monitoring, and managing planetary changes. And, over the past 50 years, increases in resolution, revisit times, swath width, and sensor capabilities across the spectrum all have improved the observations and distinctions discernible from space.

On account of the growing kinds and numbers of imaging satellites, as well as terrestrial, aerial, maritime, and submersible geospatial sensors, answering questions about climate change and the relative warmth of the planet has created numerous opportunities for geospatial envisioning. The maritime opportunities created by Saildrone, provide one example. Saildrone came about when someone envisioned how to combine the very old maritime propulsion technology of a sail with 20th century GPS technology and 21st century solar-powered sensors. Other examples of envisioning related to global warming include the combination of modern precision location technology with the 18th- and 19th-century technologies lighter-than-air balloons and ultra-light aircraft and gliders. The natural propulsion by air currents enables these platforms to loiter and to travel at a pace that allows more accurate and longer duration measurements of atmospheric conditions.

The analytic challenges of climate change afford more opportunities for scientific envisioning that combine old and new technologies to provide new sources of geospatial information and geospatial intelligence. The examples of imaginative drone use are studies of volcanic activity or the effects of nuclear accidents. The scientific community has adapted military drone technology for nonmilitary purposes, and drones are being flown to gather intelligence about air and water quality in irradiated or volcanic terrain judged unsafe for humans. Both aerial and robotic terrestrial drones have been used to assess the amounts and effects of uncontrolled radioactivity after the Fukashima Reactor explosion following the 2011 tsunami.[4]

Measuring the envisioning results of new geospatial technologies has not been difficult as they have led to new and more rapid discoveries. After the earthquake under the Pacific Ocean that caused the tsunami that struck the Dai Ichi reactors at Fukashima, geospatial technology rapidly discovered the seafloor changes in the Pacific caused by the earthquake. Due to this technology, the United States Geological Survey (USGS) was able to communicate rapidly the location and extent of these changes, and how they would alter the safety of navigation in the region.

Similar terrestrial discoveries can be measured precisely at a smaller scale. After the USGS seismologists reported evidence of North Korean nuclear testing in 2016, 38 North analysts and image scientists were able to discover and measure, using synthetic aperture radar change detection, the exact location and extent of surface changes on the North Korean mountain caused by the underground test.[5] Similar studies using SAR are being done after major earthquakes and during increased volcanic activity to discover changes in terrain elevation and location.

In the air, like the sea and land, geospatial intelligence and technology have helped discover the extent and effect of natural phenomenon. After the Icelandic eruption that created a new volcano in 2010, geospatial analysis conducted by USGS, the Icelandic Meteorological Office, and Eurocontrol discovered and plotted the physical location of the ash plume created by the volcano in the atmosphere. These organizations published geospatial intelligence that enabled commercial airliners to reroute or cancel flights on the basis of this information.[6] This intelligence kept the airlines from the increased risk to equipment and flight safety.

One byproduct of commercial geospatial intelligence is that it has accelerated the pace of scientific discovery. When the US government controlled the geospatial intelligence technology in the 1970s and 1980s, the scientific community knew that the government had employed additional remote sensing capability, but it did not get to see the 1960–1972 CORONA film-return imagery until it was declassified in 1995, under the prodding of the Clinton administration, specifically Al Gore's reinventing government initiative.

While many of the first-generation commercial imagery sensors and platforms had been engineered to provide information and intelligence for national security questions, very quickly after being launched, the commercial satellite companies found a secondary market in helping nations rescue, respond, and recover from natural disasters. Their imagery did not have to be declassified, and it could be disseminated more quickly to a wider audience. This market expanded after numerous analytic organizations began to use this imagery and other geospatial data sources to comprehend various energy, pollution, and environmental effects questions. The commercial imagery providers could collect with sufficient frequency or periodicity so that these companies and organizations were able to comprehend the effects of human activities in some places on the planet. Yet, in spite of the innovative successes of geospatial technology over the past 20 years, the challenges of recording and comprehending global warming/climate change remain daunting.

At least 20 years of geospatial analysis exists on the issue of global warming, but it does not exist coherently or accessibly. The most common parts and the greatest amount of this work have been done with ESRI software. Yet, unlike the global public health community did with COVID-19, the global scientific community has not agreed upon a single sharable database for recording these findings. No universally accepted agreements exist about shared formats among all the organizations working on global warming. While this situation may not be typical in the scientific community, it seems very familiar to veterans of the US Intelligence Community.

During the Cold War, in the US, even with only one source of satellite information, the same spot on earth might have as many as four recorded coordinates. Prior to the accepted use of GPS in the US Intelligence Community, the Department of State kept official gazetteers with approved place names and associated coordinates. The National Reconnaissance Office kept a separate secure data base used for tasking satellites. The National Photographic Interpretation Center had another separate classified database where analysts plotted, by hand from hard copy film, the new targets they found. And finally, the Defense Intelligence Agency had a fourth database as it had organizational responsibility in the Defense Department for target names. Naturally inside one intelligence community, four data bases caused a significant bureaucratic problem with contradictory data, particularly the specificity of coordinates.

For the 21st century issue of global warming, the recording problem is daunting for all nations using geospatial intelligence to track changes. There is no agreed-upon global tasking plan. No individual agency, government, or inter-governmental organization can claim success at organizing the recording of information and intelligence against this challenge. Many nations record observations and information related to geospatial discovery, but no globally shared data base nor international protocol exists to record and share this data. To achieve any success, some agreed-upon principle for recording will be required. It will have to accommodate all the nations, organizations, companies, and educational institutions that have launched or are preparing to launch small satellites with sensors. Since the end of the 20th century, those launches have numbered in thousands, and in the next decade, the expectation is that they will number in the tens of thousands.

As the Cold War had been considered an existential threat to nations, global warming may start to be envisioned as an existential threat to individuals. Perhaps a bottom-up effort will force all the global geospatial databases to coalesce to address this issue. The immense recording challenge sits in front of the more immense global analytic comprehension challenge, and until some agreed-upon format and connectivity is achieved, similar to the Johns Hopkins COVID dashboard, all working on this challenge will suffer from the lack of accepted recording standards that will impair the geospatial analysis.

Daily, reporting on climate change comes from many different sources and places. So much reporting is ongoing that those concerned or responsible for analyzing or acting in response to global warming have little or no time to reflect and think. While much is understood about global warming, the significance of new discoveries cannot always be related to prior events. For example, the National Snow and Ice Data Center in Boulder, Colorado, has been studying the patterns and amounts of Arctic and Antarctic ice cap creation every year since 1976.[7] Yet the significance of the changing rates of polar sea ice formation is not fully understood.

While the Center has improved its capability to measure changes in the ice caps more quickly, accurately, and frequently, that impressive achievement has not yet been translated into any capability to anticipate or forecast future events. In part this is a consequence of the size of the areas under consideration, and in part it is a function of inadequately understood variables such as wind patterns, storms at sea, and ocean current variations. The annual variation in the size of the ice caps is not only interesting to scientists. The naval powers of the world, as well as the maritime shippers of products and energy sources, as well as the fishermen, and the indigenous peoples in the affected regions are all directly challenged by this inadequately comprehended pattern.

Weather patterns, long studied with geospatial technology, are also not yet sufficiently comprehended. The rainfall patterns that have shrunk the rivers in the American West and in western Europe are easily measurable, but their causes and schedules remain unpredictable. Scientists believe that relations exist among these observed events, but in spite of the increased geospatial collection and analysis of the data, the desired understanding of the variable events in weather and climate has not yet reached the state where it is accurately predictable.

Yet progress continues. The early use of commercial geospatial satellites was for disaster response. This usage led the satellite operators to accelerate the pace at which analysts could receive imagery for analysis. For certain natural disasters—hurricanes, floods, and certain kinds of volcanic activity—geospatial analysis can be anticipatory in providing accurate measurements about what has changed. For other natural disasters, such as earthquakes, windstorms, tornados, grassfires, and some kinds of volcanic activity, the technology does not yet exist for detecting these events rapidly enough to predict natural events or to always be able to collect data during the actual event.

Whether it is quick enough for warning or too slow to assist in anything but recovery and limited rescue, geospatial intelligence increasingly shapes the analysis of climatic events. Currently, the collection and analysis of these events concentrate on comprehending the physical effects and how they have interrupted the human flows or society. In some cases, the analysis can be applied to functional effects. Functional analysis of these effects provides the information needed to estimate when the recovery will be complete and the social flows—potable water, food, sewage, electricity, wireless, energy, and medical care—can resume. The geospatial analysts' experience at local disasters is transferable in part to the broader areas involved in climate change studies.

NOTES

1 Adams, Henry. "The Dynamo and the Virgin," Chapter 25 in *The Education of Henry Adams* (1918). www.bartleby.com/159/25.html, viewed 27 July 2022.
2 www.cnet.com/science/space/ukraine-asks-commercial-satellite-operat ors-for-help-tracking-russian-troops/, 2 March 2022, viewed 11 October 2022.
3 "Satellite Images Show Bodies Lay in Bucha for Weeks, Despite Russian Claims," *New York Times*, 4 April 2022, www.nytimes.com/ 2022/04/04/world/europe/bucha-ukraine-bodies.html, viewed 27 July 2022.
4 Madrigal, Alexis, C. "Inside the Drone Missions to Fukashima," *The Atlantic*, 28 April 2011, www.theatlantic.com/technology/archive/ 2011/04/inside-the-drone-missions-to-fukushima/237981/, viewed 27 July 2022, and Peeva, Alexandra, "Now Available: New Drone Technology for Radiological Monitoring in Emergency Situations, 1 February 2021," International Atomic Energy Agency, www.iaea.org/ newscenter/news/now-available-new-drone-technology-for-radiologi cal-monitoring-in-emergency-situations, viewed 27 July 2021.
5 Pabian, Frank, Joseph S. Bermudez, Jr., and Jack Liu, "North Korea's Punggye-ri Nuclear Test Site: Satellite Imagery Shows Post-Test Effects and New Activity in Alternate Tunnel Portal Areas," 12 September 2017, 38 North, www.38north.org/2017/09/punggye091217/, viewed 27 July 2022.
6 Ritter, Malcolm, "Ash Cloud from Iceland Volcano Shuts Down Air Traffic," *Christian Science Monitor*, 15 April 2010, www.csmonitor. com/From-the-news-wires/2010/0415/Ash-cloud-from-Iceland-volcano-shuts-down-air-traffic, viewed 27 July 2022.
7 https://nsidc.org/, viewed 27 July 2022.

BIBLIOGRAPHY

Adams, Henry. "The Dynamo and the Virgin," Chapter 25 in *The Education of Henry Adams* (1918). www.bartleby.com/159/25.html.

www.cnet.com/science/space/ukraine-asks-commercial-satellite-operat ors-for-help-tracking-russian-troops/, 2 March 2022, viewed 11 October 2022.

Madrigal, Alexis, C. "Inside the Drone Missions to Fukashima," *The Atlantic*, 28 April 2011, https://nsidc.org/.

Pabian, Frank, Joseph S. Bermudez, Jr., and Jack Liu, "North Korea's Punggye-ri Nuclear Test Site: Satellite Imagery Shows Post-Test Effects and New Activity in Alternate Tunnel Portal Areas," 12 September 2017, 38 North, www.38north.org/2017/09/punggye091 217/, viewed 27 July 2022.

Peeva, Alexandra, "Now Available: New Drone Technology for Radiological Monitoring in Emergency Situations, 1 February 2021," International Atomic Energy Agency, www.iaea.org/new scenter/news/now-available-new-drone-technology-for-radiological-monitoring-in-emergency-situations, viewed 27 July 2021.

Ritter, Malcolm, "Ash Cloud from Iceland Volcano Shuts Down Air Traffic," *Christian Science Monitor*, 15 April 2010, www.csmonitor. com/From-the-news-wires/2010/0415/Ash-cloud-from-Iceland-volc ano-shuts-down-air-traffic, viewed 27 July 2022.

"Satellite Images Show Bodies lay in Bucha for weeks, despite Russian Claims," *New York Times*, 4 April 2022, www.nytimes.com/2022/ 04/04/world/europe/bucha-ukraine-bodies.html.

CHAPTER **8**

Future Issues for GEOINT

INTRODUCTION

The continued development of this new kind of intelligence will require both traditional human habits and changes in modern technology. The relation between the technology and the human response to it will develop and mesh in a changing dynamic. New digital geospatial technology will be combined with new human envisioning to bring more accuracy to precision measurement from space and more penetrating insights from analytic minds. How this particular combination of human intellect and spatial technology continues to develop is worth examining. This chapter examines the issues that test Human Intelligence and examines trends in the changing geospatial technology.

Many new issues shape the immediate future of geospatial intelligence. The human capacity for attention may govern the utility of the increasing volume of geospatial data and the accelerating rate at which it is being generated. Increasingly, the collection of this spatial data intrudes on human privacy. While the data trends bode well for geospatial discovery, they bode less well for the probability of geospatial comprehension, and geospatial tracking. The amount and focus of human attention will shape and constrain these trends.

The global access to Google Earth, enabling anyone with a digital screen to see anywhere from above far better than Da Vinci, also means that the current volume of geospatial data has exceeded the human capacity to scan it all, much less study it. This phenomenon began to influence the American intelligence community intermittently after the introduction of the KH-9 HEXAGON satellite in the 1970s,[1] and is now common throughout the domains of government and commercial geospatial intelligence.

The extraordinary capability of GPS to bring accurate locational data to everyone with digital connectivity on the planet has created a great

DOI: 10.4324/9781003436836-9

number of new uses for spatial information. Some of these uses cause anxiety and fear. Very few human occupations and activities have not been changed by precise locational data. Yet, this power to identify locations ubiquitously, quickly, accurately, and repeatedly challenges and sometimes threatens individuals and groups. Some of these individuals and groups wish to act without being observed or surveilled. Others dislike sharing their data involuntarily with unregulated organizations that use it without permission. Data spoofing is one response to this trend, as, in some regions, are increasing social and legal constraints on geospatial activities. Geospatial intelligence, once the provenance of governments alone, now can be accomplished by individuals, and few of its related digital activities are regulated.

The geospatial technology that created this data is trending toward smaller hardware, greater frequency of collection, more extensive imagery coverage, more rapid dissemination of data, and more algorithmic identifications of man-made objects. We are not yet in a world of ubiquitous sensors, but we are in a world where every individual should expect multiple times every day to be imaged from the ground or overhead. Throughout human history, nations, groups, and organizations have used intelligence to look to the future to diminish external risks and increase the accuracy of their identifications, warnings, estimates, and assessments. Geospatial intelligence—the seeing intelligence—quite reasonably ought to look better and farther into the future, and into the issues that impede and impel its progress.

HUMAN SOFTWARE DEVELOPMENT

One year after the 1971 launch of the first digital remote sensing satellite, LANDSAT, Herbert Simon was the first to write about the relation between the volume of data and the human capability to make sense of it.[2] Simon pointed out that in an age of too much information, the scarce resource becomes human attention. In the geospatial intelligence profession, the economics of attention has reversed its polarity over the past 30 years, yet most geospatial organizations have not fully changed their operations in response to this shift. Satellite imagery and geospatial data formerly were very expensive. The histories of the Cold War imagery intelligence and cartographic organizations describe the challenges of obtaining aircraft and satellite imagery and the sparse statistical sampling of parts of the planet from the 1950s through the 1990s.[3] By the 21st century, the scarcity in the economics of geospatial intelligence had shifted from collection to analytic attention.

The current shortage of human attention when confronted with a rapid increase in the volume of information at an unprecedented velocity is not unique. In the Western world, it happened during the Renaissance. While many individuals and organizations then had their beliefs shattered, the Renaissance incorporated new technologies to cope with the volume of information. Its new technologies—printing, libraries, increasing public literacy—were text-based as at that time human eyesight was the dominant sensor. Initially these Renaissance technologies disrupted conventional thinking and educational practices, but later they became commonplace.[4] Those who first learned to cope with and then exploit the new Renaissance technologies discovered the time to think about what these technologies enabled them to discover. Some of this group could envision and create more new technologies that enabled them to continue to learn and make new discoveries.

In considering the 21st century attention challenge, history tells us that it is reasonable to expect that some current geospatial analysts will figure out how to cope with technological challenges, and of that number some will figure out how to envision and discover new forms of geospatial intelligence and technology.

Those seeking the tools that address the scarcity of human attention should search human software. Four uniquely human characteristics—memory, curiosity, clarity, and insight—can be considered as both the current constraints on the growth of geospatial intelligence and the future keys to envisioning and discovery. Computer memory, once a severe and expensive technical constraint, is now so cheap and plentiful as to be thrown away in disposable objects without recycling. Yet, unlike human memory, digital memory remains constrained to what can be programmatically and mathematically entered in advance. In contrast, in an instant, human memory can be triggered metaphorically, synesthetically, or associatively in multiple ways, not all of which are fully understood or explainable.[5]

Curiosity is inexplicable. Young children are full of it, and some adults retain it. Some of these adults, if trained as geospatial intelligence analysts, combine their curiosity with a good memory and the ability to discover and communicate intelligence ahead of events. Some humans are able to look at tens of thousands of images and retain information that most of us cannot detect or hold. And a few have what may be the scarcest human commodity—insight.

This short book began with the attribute of envisioning, and one component of envisioning is insight. Little academic work in English has been written on this subject as it relates to the practice of intelligence, save the work of Gary Klein and Adrian Wolfberg[6] but human insight seems to have been distributed unevenly. In the progression of cartography, photointerpretation, imagery analysis, and geospatial intelligence, insight has been possessed by a few individuals who made great advances—Leonardo da Vinci, John Snow, James Allen, John Moore-Barbizon, Paul-Louis Weiller, Edward Steichen, Edwin (Din) Land, Art

Lundahl, Jack Dangermond, John Rohlf, and Ensheng (Frank) Dong. Their contributions illustrate the power of human creativity, and they also provide hope that future human insights will continue to develop this profession.

WORKING THE SEAMS

Human software characteristics—memory, curiosity, clarity, and insight—cannot be directly applied to geospatial intelligence problems. But their indirect application is at least as important as the computer software developments in feature recognition algorithms. The growing computational power of deep learning, quantum computing, and generative adversarial networks records and recovers that which is known and imaged faster than ever before. When these technologies are woven together, they can provide a digital hindsight that will be more reliable, recoverable, and revisable than any number of human analysts. Every day the attributes of these technologies become more essential. Yet these technologies represent the rearview mirror of geospatial intelligence.

Geospatial intelligence requires Human Intelligence to provide analytic foresight to complement digital and algorithmic hindsight. Daily, human analysts address the unknown and the unforeseen with incomplete observations, detections, and intuitions. The human abilities of memory, curiosity, clarity, and insight—the foundations of envisioning—are at least as necessary to geospatial intelligence as the technology. They enable human analysts to make discoveries, detect similarities, recognize paradoxes, and connect initially dissimilar observations in new and useful ways. They enable humans to use analogy and metaphor to make connections and relations that sensors cannot. These uniquely human abilities are the other essential elements of the geospatial genome, and much academic research remains to be done on exploring the relations between geospatial technology and the human imagination.

And as geospatial intelligence develops and changes, future patterns of envisioning and discovery will enable comprehending, recording, and tracking of changes. As Leonardo da Vinci demonstrated first and future geospatial analysts will continue to do, geospatial intelligence will continue to create the potential for mankind to see the world differently.

ATTENTION DEFICIT REORDER

The final future trend is the high rate of consumption that geospatial intelligence requires. It demands physical energy from the sensors and empirical and intellectual attention from analysts. Analytic managers will have to decide how much intellectual energy to expend against long-standing and current questions that require keeping up, and how much of the same scarce resource to anticipate, address, and possibly answer future questions that require staying ahead of today's issues. Currently analytic managers in the US Intelligence Community have to accomplish this challenge without a replacement for the 20th century Cold War accountancy collection and analysis model.

With increasingly ubiquitous GEOINT collection, what are governments and organizations to do? Existing formal burden-sharing approaches include the National System of Geospatial Intelligence, the bureaucratic structure in the US Intelligence Community by which the National Geospatial-Intelligence Agency (NGA) works jointly with other components of the US Intelligence Community and the Department of Defense. NGA also works jointly with federal and civil agencies through the Civil Applications Committee, part of the US Department of the Interior. NGA also partners with other Allied nations through the Allied System of Geospatial Intelligence which comprises the UK, Canada, Australia, New Zealand, and the US.[7] NATO has also created a shared definition of geospatial intelligence so that its member nations can facilitate their cooperative efforts.[8] The Ukraine War of 2022 has accelerated this information sharing in practice, if not in bureaucratic agreements. Yet none of these negotiated arrangements has convinced any of the participants that the bureaucratic nirvana of sufficient geospatial resources has been reached.

While bureaucratically impressive, government cooperation among all the agencies, commands, services, and departments engaged in geospatial intelligence remains operationally inadequate. It is impressive as all government agencies recognize the growth of this new form of intelligence, and it is inadequate as the growth in the numbers of international and commercial geospatial intelligence sensors and platforms far exceeds the collective analytic capacity of the governmental organizations.

Additional reordering of the federal geospatial community is likely to become necessary, and some organizational precedents exist. Temporary partnership between commercial entities and governmental organizations is one example. The 1957 Jam Session model is an exemplar. Another approach would be the professional-amateur model as Skytruth successfully used in its directed search for fracking sites in Ohio and Pennsylvania. It is likely that these future joint ventures will include feature recognition algorithms and open-source crowdsourcing on one end of the continuum, and more exquisite classified "tradecraft" on the other end. The important middle ground is not yet defined, but a reordering of analytic processes is needed to deal with the attention shortage. The middle ground will likely require some mix of human analysts and analytic software.[9]

Another form of partnership is underway. In a manifestation of the crowdsourcing effort and in response to US Congressionally directed action, the NGA initiated in 2021 the Tearline program (tearline.mil).[10] The Tearline program aims to make connections between academic and unclassified GEOINT efforts and analytic efforts inside the US Intelligence Community. Tearline also aims to make exemplary academic geospatial analysis easily accessible to the intelligence community and to distribute academic unclassified geospatial analysis to the citizenry.[11] Where this new effort will lead is uncertain, but it bears watching and perhaps consideration by other governments developing their own geospatial intelligence efforts. The pace of technological changes will force essential human and organizational changes.

Eight current technical trends are changing in geospatial intelligence: Miniaturization, Global GEOINT, Commoditization, Collection as Analysis, Precision, Acceleration, Continuous Ubiquity, and Mobile Geospatial Intelligence.

MINIATURIZATION

Throughout the Cold War, the operating premise for the US and the USSR in space was simple: size matters. Increased mass was a goal and an operating principle for ballistic missiles, nuclear testing, and imaging satellites. Both nations increased the size and launch capacity of their ballistic missiles from 1957 until 1975 when the Soviet Union deployed the SS-18 and the US deployed the LGM-118 Peacekeeper in 1985.[12]

In nuclear testing, both the US and the USSR increased the size and lethality of weapons until the 1954 US Castle Bravo Test in Bikini and the 1961 USSR Tsar Bomba test at Novaya Zemlaya.[13,14] The unanticipated radioactive dispersal effects of these tests led to developments that resulted in both nations signing the 1963 nuclear non-atmospheric test ban, which banned future aerial and underwater testing.[15] This first arms control treaty, initially among the US, the UK and the USSR, paved the way for the use of photographic and imaging satellites in monitoring arms control agreements during the Cold War.

During that time, the growth in imaging satellite size and film capacity paralleled the growing concern about which missile system could achieve the greatest throw weight.[16] The last development in film-based space photography, the US KH-9 (1972–1985), carried up to a ton of unexposed film on each launch. This capability meant an exponential growth in the amount of film that each of the four KH-9 buckets carried back to earth for photointerpreters to look at. By 1985, at intervals, the US had more satellite imagery than it could look at in any scheduled fashion. Yet at the time the increase in photographic intelligence data did not result in an analogous increase in discovering or comprehending.

In the mid-to-late 1980s the US recognized that more information from satellites was not necessarily better information. This discovery may have begun the shift in thinking that enabled the National Reconnaissance Office (NRO) and other agencies in the US Intelligence Community, legitimately criticized by the US military after Desert Shield/Desert Storm, to respond so swiftly 12 years later in Operation Iraqi Freedom. In the interval between these two Iraq wars, the NRO invested in the digital modernization and installation of ground stations, and DoD began to use civilian digital communication architectures. These measures improved the US ability to move digital satellite data around the planet and provided much more rapid intelligence dissemination to the US and coalition forces in the Middle East.

But the most significant and dramatic change in geospatial technology—miniaturization—was not introduced by government defense or intelligence agencies. The academic and private sector developed small satellites, micro-satellites, and nano-satellites in the first decade of the 21st century. By the second decade, a commercial small imaging satellite company, Planet Labs, had launched and orbited a constellation of smallsats flying in formation that could image the entire land mass of the earth every day. The entire Planet constellation weighs less than a tenth of one KH-9 film-return satellite.[17]

Reduced launch and orbital operating costs accompanied the miniaturization of geospatial hardware. In 2021, of all imaging satellites on orbit, governments operated fewer than 20%, and that percentage is dropping yearly. Miniaturization has driven down the cost of geospatial technology. The changing economics of satellite construction and space launches has moved GEOINT from the government to the private sector, and partially into the nonprofit and educational sectors.

Globally, educational institutions, at the graduate, university, and high school levels, fly about 34% of the existing small- and nano-satellites.[18] As launch costs and satellite mass continue to shrink and organizational interest increases, this trend will grow. More and more nonprofit organizations and privately funded advocacy groups will finance and operate their own imaging constellations to focus collection to provide information and answers to questions they care about.[19]

Miniaturization radically altered geospatial analysis and collection management. Geospatial analysis in support of civilian disaster support or military operations routinely works with increased amounts of imagery coverage either in the field or analyzed elsewhere and transmitted to the field. For collection planning, a commercial app, Spymesat, exists for anyone with a smart phone or computer. It shows over a 12-hour period, when coordinates are entered, which commercial satellites could image that point on the earth with a specific sensor at a specific time and at a specific cost. These advances in miniaturization greatly improve the input before geospatial analysis and the output after the analysis. Yet for all their potential they also have a dark side.

The malign use of miniaturized geospatial technology has already been demonstrated. In the fight to destroy the Islamic State in Iraq and in Syria (ISIS) caliphate in northern Syria and northern Iraq in 2017, ISIS armed small drones with explosive charges and used commercial geospatial technology to create a poor-man's precision weapon.[20] The legal and ethical constraints that shape US and Allied targeting are not considered by terrorist organizations. US analysis of the 2021 Iranian ballistic missile attack on US Forces at Al Asad airbase in Iraq points to the Iranians launching their missiles after their review of that day's most recently collected commercial imagery.[21]

The Spymesat handheld app for planning commercial collection management can also be used, in the inverse, to schedule activities at any location so as to deny their observation from commercial imagery collection.[22] With very little effort, anyone can identify when any location will and will not be in the field of view of upcoming satellite collection. Electronic tools for satellite detection date back to the Soviet Satellite Warning system in the Cold War.[23] Now, anyone can download that same capability into a mobile phone for free. Any group can become aware of the schedules of global surveillance of their future planning, rehearsal, logistical preparations, and movements.

GLOBAL GEOINT

Since 2000, geospatial satellite constellations have become cheaper, smaller, more numerous, and more international. Formerly the exclusive domain of powerful nations—the US, the USSR, China, France, Israel, and India—in 2022 satellite imaging programs spread to 74 countries and that number will continue to grow.[24] In the 21st century, having a national remote sensing program seems to be considered a modern indicator of national prestige, much like national airlines were after World War II. The reduced cost and greater availability of missile launches, the smaller size of imaging satellites, and the reduced cost of computer processing and digital communications have made these changes possible.

Private corporations such as Planet, Maxar, Airbus, Spire, Hawkeye 360, Capella, Iceye, Blacksky, Satellogic, Synspective, and others fly constellations of imaging satellites. The proliferation of global small satellite companies will likely follow the tradition of other radically new technologies such as aircraft and automobiles. The companies that produce them, many at first, will merge over time. The output from all these satellites—data, processed imagery, and analytic products—is now being analyzed and processed by some governments and international agencies as well as commercial and nonprofit geospatial analytic organizations such as Descartes Labs, Ursa, All-Source Analysis, 38North, Skytruth, and Global Fishing Watch. So far, this free market use of private geospatial data and analysis has been benign.

These companies have provided greater accuracy in global commodities markets in agriculture (Descartes Labs) and petroleum storage (Ursa). Some organizations also informed the world about and advocated against the activities of totalitarian regimes such as Sudan (Satellite Sentinel Project), and North Korea (38North). Multispectral constellations are beginning to measure methane and CO_2 emissions globally.[25] Yet, for all the geospatial information and analysis produced by these systems and organizations, the market is not clearly defined. In some areas of the world, regulatory agencies exist and emissions standards are enforced. But the same cannot be said for all parts of Asia and Africa. Yet every good geospatial action in space can have an equal and opposite geospatial reaction on earth.

For example, Global Fishing Watch, a subsidiary of Skytruth, pioneered the use of geospatial technology to analyze global fishing patterns. They accomplished this by tracking automatic identification sensor (AIS) ship-location detectors to identify vessels fishing in prohibited waters. The fishing vessels that turned off their AIS sensors as they approached restricted areas were clearly trying to avoid detection, but their actions made them detectable and identifiable by Global Fishing Watch. By so doing, this small organization served notice on the world's fishing fleets that their movements—open and surreptitious—were being tracked.

But as part of their monitoring of commercial fishing fleets, Global Fishing Watch discovered that GPS and AIS signals associated with Russian cargo ships were being plotted inland in central California. Further analysis revealed that these Russian signals had been spoofed so as to obscure their current locations in the Black Sea and the Mediterranean and to conceal their future destinations.[26] Groups intent on obscuring their locations and destinations have recognized the global access to geospatial analysis, and these groups actively practice denial and deception measures against small nonprofit commercial geospatial organizations, as they formerly had against nations. However, only failed deception efforts are noticed, and it is to be expected that more sophisticated spoofing is being and will continue to be developed in response to commercial sensors.

The rate of future algorithm development for detecting features on the increasing volume of commercial digital imagery is hard to measure. Yet a sense of the scale of improvement is measurable. The first feature recognition algorithm, Frank Rosenblatt's Mark I Perceptron, in 1958, combined 400 sensors with 512 potentiometers to achieve 204,800 possible parameters or 2.048×10^5. The 2012 Microsoft algorithm that won the ImageNet competition had 60 million parameters or 6×10^7. In 2020, a linguistic neural net algorithm, GPT-3,[27] had 175 billion parameters or 1.75×10^9. While the GPT-3 is not a feature recognition algorithm, it points toward their future.

The two orders-of-magnitude growth in the number of parameters between the Perceptron and the ImageNet algorithm took 54 years, but the more recent growth of two orders-of-magnitude took only 8 years. This rate of change is not likely to diminish. As algorithms learn faster, become more accurate, and more commonplace, they will expand their analytic applications. Yet algorithms can detect only that which is already known.

COMMODITIZATION

The expectation of continued technological improvement and the growth in access to geospatial information and intelligence is changing how most people think about these topics. Geospatial intelligence has turned into a commodity. What was once esoteric and spooky technology from the world of espionage has become familiar. After the US invaded Iraq in 2003, most people have become so accustomed to seeing satellite imagery in media or websites that it has become so common as to be unremarkable. The familiarity that geospatial intelligence has engendered in society has bred, if not contempt, then a routine expectation. Perhaps the best illustration of this routine response was exhibited by former President Donald Trump in August 2019, when he used a cell phone to photograph a Top Secret US government image of a failed Iranian missile test.[28]

The conversion of a highly classified intelligence report into fodder for a social media taunt illustrates the commoditization of geospatial data and how official perception by some changed geospatial intelligence from a context of secrecy to a means of public mockery. The approach to treating geospatial intelligence as though it will always be available has grown increasingly common both in and out of government. While freely available commercial geospatial intelligence has made the keeping of secrets harder for those who would rather work in secret, such as Iranian missile engineers, it has also made the discovery of secrets harder for those governmental geospatial analysts who use this technology to unearth potentially threatening developments and to manage future risks.

Geospatial data is compiling at a rate comparable to the melting of the icecaps. And, like all the reporting about climate change and ocean warming, human attention to hourly and daily information relating to global climate or the increase in geospatial data has a finite limit. The Ukrainian use of commercial satellite imagery for media had made their public releases routine. Yet, the human response to information overload is not always thoughtful. Reporting on Ukraine has gotten to the point that optical imagery over Ukraine is more noteworthy in the media reporting by its absence on cloudy days. These trends of increased access and diminished attention deserve consideration as they are shaping how geospatial analysts and geospatial organizations address their professional challenges.

COLLECTION AS ANALYSIS

Historically, in many geospatial organizations collection management has been the servant of analysis. For a long time, collection managers have been considered the process stewards who obtained mapping data or imagery for analysts or cartographers. During the Cold War, accounting served as the intellectual model for collection management in the US Intelligence Community.[29] James Reber, who in the 1950s and 1960s had to reconcile and rank all the potential intelligence targets against the limited collection capabilities of the U-2 and the CORONA photo-satellites, devised the measures of success for imagery collection in a time of scarcity. He created the accountable search plan for the Soviet Union and instituted as a measure of success the ratio of the measures of images acquired in relation to images required to discover, comprehend, or track activity at a target or targets.

When imagery was scarce, Reber's accountancy measures worked very successfully. But the explosive growth in collection brought about sporadically by the KH-9, increasingly by the KH-11 in the 1970s and 1980s, and routinely by commercial digital imagery in the 21st century requires a new model for collection management. The new model has to account for five criteria:

- Swath width, or the area that an individual satellite could collect at one time.
- Resolution, or the interpretability of the product of the sensor. This is measured in ground sampling distance.
- Periodicity or the frequency of required collection. This can also be thought of as the percent of persistence, or what amount of the 86,400 seconds in a day need to be imaged over a particular target to answer a particular question.
- Cost, as in the 21st century, most geospatial collectors are commercial and charge for their data, access, images, products, and analysis.

Finally,

- Analytic attention, as too much geospatial data and imagery are collected for the number of analysts to look at.

The first three factors have informed collection management since World War I, but in this era of exponential growth in imaging satellites and imaging drones the last two factors—attention and cost—have become equally important.

In the 21st century commercial imagery, providers initially staked their business plans on one of these five factors. Planet Labs focuses on daily area collection; Maxar focuses on high-resolution imagery, and synthetic aperture radar companies like Capella and Iceye focus on access during nighttime and cloud cover when optical systems cannot image. Drone manufacturers have used video technology to increase their periodicity and persistence. And for these and all the other geospatial companies, cost and analytic attention affect their business successes.

As more and more imaging satellite companies in more and more countries come on orbit, cost becomes a challenge for commercial satellite companies as their margins shrink. And for the customers of these companies, the volume of geospatial information increasingly challenges the available attention of all the analysts. Geospatial intelligence now requires measuring and monitoring the available amount of analytic attention. To avoid strategic surprise, new units of measure will have to be derived for level-of-effort programs. For analytic managers, they will have to decide how much intellectual energy to expend against the current intelligence questions and to keep up with current issues, and where to find the human attention resources to anticipate and answer future intelligence questions that require staying ahead of current issues.

Success for analytic managers will mean relying on the analytic and probabilistic expertise of the collection managers. The current geospatial collection environment requires a different mathematical discipline—probability instead of accounting. In this world of copious daily geospatial data, a sampling strategy will need to be created, maintained, and continuously revised so as to accommodate discoveries by geospatial analysts at a location. This strategy will be needed so the analysts can sample this location at a rate faster than the pace of change. Collection management in this era will require understanding of potential risk as much as an understanding of managing analysis. Twenty-first century geospatial collection will require an increased understanding of probability, and the geospatial analysts will have to supply the necessary attention.

The world will continue to have unsettled places. And in these unsettled places, the intelligence demands for greater swath width,

multiple sensors, increased resolution, and more persistent surveillance will converge and create future shortages of available collection.
North Korea is currently such an unsettled place. For these unsettled
locations, accounting as an intelligence collection model will not go
away entirely, but it will have to be applied in combination with a probabilistic mindset.

PRECISION

Geospatial precision has brought much good to humanity in accelerating
disaster responses, identifying individuals or groups in distress, and enabling the targeting and tracking of nefarious groups and individuals. Yet
criminals have demonstrated their ability to use precision locational data.
Nefarious groups, such as human traffickers, drug smugglers, terrorists,
and weapons proliferators, understand geospatial technology and work
very hard to undercut its effects. Russian ships spoofing AIS signals, a tool
for maritime rescue, to evade discovery or break the law, is one example.

In addition to locational precision, the image quality of terrestrial
digital sensors is improving rapidly. Most people on the planet now carry
multiple sensors that broadcast locational and other information—cell
phones, smart watches, hearing aids, pacemakers, and smart glasses are
only a few of these devices. The internet of mobile things relies on this
technology that enables more and more of the human beings on the
planet to create an accurate chronological and time-stamped map of their
whereabouts.

The positive side of precision location provides governments,
organizations, and people more timely, discrete, and socially useful information. This technology has transformed the logistics of supply chains,
marketplace options, and the quality of services provided by societies and
governments. It made possible global tracking of the spread of COVID-19
as well as the vaccine dissemination that is shrinking the effects of the pandemic. Location-based thinking fueled the growth in large social-media
companies such as Facebook, Instagram, LinkedIn, Spotify, Snapchat, and
others. But the free service comes with a proviso.

Social media companies thrive by inhaling the digital exhaust of
their members. The members surrender the precise temporal and spatial pattern of their lives, wittingly or unwittingly, to these companies. In
these companies, sophisticated location-based algorithms underpin other
queries that record and locate credit card purchases, travel, electronic
communications, and most importantly, each of our digital preferences
and movements. The electronic constants in all digitally recorded human
transactions—fiscal, social. recreational, governmental, charitable, medical, and occupational—are location and time.

In most cases volunteering our individual locational information is neutral or beneficial. Yet it can result in malign or evil outcomes. The tacit agreement of individuals to be photographed, recorded, or digitally tracked has led to the rise of surveillance societies and states. While this may be a general social good, not every outcome it creates is benign.

When a common location technology is considered, the digital technology that automatically records highway and bridge tolls and fares, the social benefits jump out. This technology provides convenience for the individual, information that allows better route planning, easier access, improved traffic management, better maintenance for the infrastructure, predictable revenues for governments, information that leads to highway safety, and a more economical and efficient highway system.

Yet the same location-based systems can be spoofed, avoided, or manipulated by those who use geospatial technology with malign intent. In spite of all the data gathered about transportation infrastructure and patterns, these systems have not eliminated international or interstate human trafficking, illicit drug and weapons shipments, car thefts, or other vehicle-based crimes.

ACCELERATION

The rates of sensor deployment on earth and in space, unmanned aerial vehicle (drone) deployments, and improvement in feature recognition algorithms and computer systems are all accelerating. This trend will not diminish. The first two developments cover all locations where human activity happens, and the last increases the potential amount of time during any day when an activity can be detected.

In February 2021, scientists linked the radio telescopes on earth to act jointly as one terrestrial synthetic aperture radar. They used this new creation to image a swath of the moon at a distance of 187,000 miles at 5-meter resolution.[30] This resolution is about eight or nine times better than the NRO's Quill proof-of-concept radar experiment of 1964[31] but over a distance more than 1,200 times greater. Much like the scientists at the Johns Hopkins Applied Physics Lab who tracked Sputnik I in 1957 and recognized that signals travelling in the opposite direction could also locate objects on earth, it is not inconceivable that all the multiple smallsat SAR sensors orbiting earth may, at some future date, be linked to create a new way of envisioning geospatial intelligence on this planet.

The current generation of cell phones advertises spatially registered digital imagery at a quality formerly obtainable only in cinematography. And the multispectral and hyperspectral sensors that are being launched on smallsat and drone platforms will provide better resolution and more discrete bands along the electromagnetic spectrum. These sensors will provide global information about gas emissions, air pollution, and botanical changes as well as detect the presence of chemical signatures for explosive components.

The rate of algorithm development for detecting features on digital imagery is harder to measure. The creators of most feature recognition algorithms consider them as proprietary information. While comparative data are hard to come by, the results are continually being advertised. Yet, some of the malign effects of these algorithms have already been identified. China may be the most advanced society that employs facial recognition algorithms, but their use to identify and track individuals from certain ethnic groups raises significant privacy and ethical issues.[32]

The easiest measurement of the growth in geospatial technology is the increasing number of small satellites announced to the public upon launch. Government agencies record their successes and failures as well. As every launch is posted on the net live, far more current satellite news is available. The bright side of acceleration is the ease of access to outer space. Geospatial data and imagery are becoming increasingly cheaper to access for anyone with an Internet connection. What is not currently free is current or ubiquitous coverage. Over the past few years, after every natural disaster, imagery providers have made free geospatial data available as a public service. After every disaster GIS providers make their products free for use by disaster responders. Google Earth and Google Maps make ubiquitous coverage available for free, but sometimes the coverage is not current nor frequently updated at a known schedule on a global basis. Yet digital data is trending toward more data being made available more frequently. On every continent, along with the accelerated technologies, the demand for more geospatial intelligence is also accelerating.

CONTINUOUS UBIQUITY

While the continuing miniaturization of geospatial technology warrants following, the costs of continual attention have to be tallied. The trends in geospatial technology are locally, nationally, and globally good. No other technology can discover, record, and enable the response, reporting, and research on changes to the environment or human social order. Individuals, groups, organizations, military forces, and governments that formerly acted without fear of discovery now learn that anyone can discover activities anywhere. Yet, the benefits of envisioning, discovering, recording, comprehending, and tracking come with a price.

The price, both positive and negative, is human attention, and no plan exists for rationing human attention at any level of government—local, tribal, state, federal, national, or international. For geospatial intelligence communities, constant vigilance about the positive and negative implications of these trends will be required. Offensively, vigilance will be required to do good and become more aware of the growing number of users of geospatial analysis throughout the globe. Defensively, it will be incumbent on the growing number of geospatial communities to monitor the effects of increasing geospatial technology on individual rights and collective security. Those new geospatial applications will have to be reviewed, and the human, network, and national costs to undertake this review will need to be calculated routinely as new subjects for geospatial intelligence arise.

For example, the increased volume of increasingly precise imagery impairs warning. Being able to see more of the world more often and at better resolution does not translate into an increased ability to warn of more or more subtle immanent threats. Effective warning relies on defined indicators and awareness of norms of expected activity. Comprehending and identifying indicators and analyzing and researching the frequency of observations to define a norm of activity require sustained attention and research over time.

While geospatial data has become more and more ubiquitous, two negative trends are rising—automated digital surveillance and diminished human attention. As knowledge about imaging satellite operations has devolved from governments to commercial organizations and nonprofits, large and small, awareness of the need to avoid surveillance technology has increased commensurately among those who threaten existing societies and governments.

Geospatial data is compiling at an exponential rate. Like the reporting about climate change and ocean warming, the finite limits of human attention to hourly and daily information relating to global climate change or the increase in geospatial data have a limit. Often, the human response to information overload is not thoughtful, and one human response is to shut down. Yet for people to warn against any threat, it is necessary to discover, define, and locate the threat in advance. It is also necessary to comprehend the threat. Tracking, much more than comprehending and discovering, requires constantly focused attention. And when attention has become the scarcest resource, individual, organizational, and environmental organizations will have to change their operations to manage this scarce resource more effectively.

MOBILE GEOSPATIAL INTELLIGENCE

Increased mobility is a trend that geospatial intelligence both drives and follows. Much good work already has gone into making the results of geospatial intelligence available globally through mobile applications. Since the Haitian Earthquake of 2010, mobile geospatial inputs, as demonstrated by Ushahidi, Open Street Map, and GeoCommons, have driven the pace and need for making rapid geospatial intelligence available through cellular telephony.[33] Yet, a visual contradiction remains. While mobile phones can process locational disaster information, a need remains to depict events, even disasters, in spatial terms. Geospatial analysts need large monitors to determine the extent of change and to provide context for the observations. These contradictory demands must be reconciled. Analysts will need to pull back to envision surrounding areas to gain context on what they might be missing, as well as to focus on the need to work through the present task of identifying what is present on the image. While the result of analysis provides valid information for the smaller monitors of mobile cellular phones, the analysts will have times when working with larger monitors will be necessary. While the geospatial product of the analysis can be seen on mobile screens, the process of analysis requires large monitors. Those able to translate spatial information between these two technologies will have a better chance of success.

In the age of open and mobile geospatial intelligence, secret and clandestine geospatial analysis will remain necessary and will be created only for a small audience. This type of geospatial analysis will continue to be produced in exclusive and secure environments to preserve for its recipients the chances for operational, intelligence, or clandestine success. As secrecy about satellites, drones, and sensors diminishes, it may become necessary for geospatial analysts to create and use a digital and visual equivalent of the cryptographers one-time-pad.

An empirical and intellectual tension exists between these points of view. First responders and military operators need information from mobile devices, and do not worry about interpretative difficulties. Geospatial analysts, who desire the best data over the target, need to be able to discern small changes over large areas to inform responders and operators about what is present and what is missing on the image. Said differently: the question for military operators and first responders is: For an immediate intelligence readout, how much thought and context are desired, and how much are required? The historical three-phase approach that began in the World War II era of photo-interpretation remains useful even in an environment saturated with new geospatial information, but it will require experience and judgment to use this approach effectively in the accelerating data-saturated environment.

The question about the volume of available information that geospatial intelligence requires can be answered. While sometimes forethought is considered a luxury, the investment of analytic resources before a crisis can accelerate and improve the needed intelligence. Yet this answer leads to an enduring question for which there is no exact answer: How much geospatial analysis is enough? Geospatial analysis is not free, or cheap, and yet, the governmental, social, economic, environmental, and defense demands for it increase steadily.

A FINAL QUESTION

The pace of geospatial acquisition and consumption continues to accelerate. Fifty years ago, in the film-return era, the lag between deliveries of buckets of film on average was slightly longer than a month. Today, hundreds of imaging satellites and tens of thousands of drones download data so frequently that the intervals between geospatial downloads have been reduced to seconds. And when all the geospatially related imaging data from cell phone images and videos is included, in some locations, there is no time when observational data is not being recorded. Increasingly, some of this data is temporarily ignored and some will never be looked at. The phenomenon of ignored geospatial data has existed for at least 50 years, but only with the recent development and improvement of feature recognition algorithms has retrospective searching through this glut of geospatial data become possible. Yet algorithms can identify only the known and distinctively shaped and not the unknown or the irregularly shaped. So, the centuries-old challenge remains: how can humans discover the unknown in all this data.

The answer to this question is unknown. But, it is very highly likely that the answer will involve:

Envisioning
Discovery
Recording
Comprehending, and
Tracking.

And as the discipline of geospatial intelligence develops and changes, the past patterns of envisioning and discovery will enable comprehending, recording, and tracking of future changes, and geospatial intelligence will continue to provide the potential for mankind to see the world differently as Leonardo da Vinci demonstrated first and future geospatial analysts will continue to do.

NOTES

1 O'Connor, *NPIC*. pp. 86–89.
2 Simon, Herbert A. "In an information-rich world, the wealth of information means a dearth of something else: a scarcity of whatever it is that information consumes. What information consumes is rather obvious: it consumes the attention of its recipients. Hence a wealth of information creates a poverty of attention and a need to allocate that

attention efficiently among the overabundance of information sources that might consume it." "Designing Organizations for an Information-Rich World," in: Martin Greenberger, *Computers, Communication, and the Public Interest* (Baltimore, MD: The Johns Hopkins Press), pp. 40–41; also Lanham, Richard A. *The Economics of Attention: Style and Substance in the Age of Information* (Chicago: Chicago University Press, 2006).

3 The previously cited official NRO histories of the satellite programs and the CIA history of the U-2 program and the HTAUTOMAT organization all address the challenges of obtaining collection as does *The Intelligence Map Program on Communist Areas 1960–1969.* (CIA, Approved for Release: 2018/02/27, C01421958).

4 Eisenstein, Elizabeth L., *The Printing Revolution in Early Modern Europe.* (Cambridge UK: Cambridge University Press, 1983), Lanham, *The Motives of Eloquence: Literary Rhetoric in the Renaissance* (New Haven, CT: Yale University Press, 1976); Haydn, Hiram. *The Counter-Renaissance* (New York, NY: Scribners, 1950).

5 Marcel Proust's *A la recherche de temps perdu* is the classic literary work that depicts sensory triggering. The most well-known English translation would be C. K. Scott Moncrieff's and Terence Kilmartin's *Remembrance of Things Past* (New York, NY: Vintage, 1982). In the field of imagery and geospatial analysis, the author at different points in his career has witnessed several events that began "I don't know what you are looking at, but it reminds me of this image of this place."

6 Klein, Gary. *Seeing What Others Don't: The Remarkable Ways that Humans Achieve Insight* (New York, NY: PublicAffairs, 2013) and Adrian Wolfberg, *In Pursuit of Insight: The Everyday Work of Intelligence Analysts Who Solve Real World Novel Problems.* Research Monograph (Washington, D.C National Intelligence University: Ann Caracristi Institute for Intelligence Research, July 2022).

7 NGA. National System for Geospatial Intelligence. *Geospatial Intelligence Basic Doctrine.* Publication 1.0, April 2018. Approved for Public Release. p. 33.

8 NATO, https://nso.nato.int/natoterm/Web.mvc. Geospatial Intelligence: Intelligence derived from the combination of geospatial information, including imagery, with other intelligence data to describe, assess, and visually depict geographically referenced activities and features on the earth. (and in French) Renseignement Geospatial: Renseignement issu de la combinaison de l'information géospatiale, dont l'imagerie, et d'autres sources de renseignement pour décrire, évaluer et représenter visuellement les activités et caractéristiques à référence géographique sur terre.

9 The word "exquisite" is a form of art in the recent history of geospatial intelligence. The author recalls hearing it for the first time during the 1990s in briefings about the Future Imagery Architecture (FIA) which was the NRO's attempt to economize in the 1990s after Desert Storm/Desert Shield. In the briefings, the word "exquisite" was synonymous with "expensive." The distinction then was made between exquisite imagery (government supplied) and commercial imagery. More recently, the distinction changed to "exquisite tradecraft," again meaning government supplied, and less exquisite or contractor-supplied tradecraft (presumably cheaper).

10 www.tearline.mil/ viewed 15 January 2022.

11 In the spirit of full disclosure, a number of students in the Johns Hopkins Geospatial Intelligence degree program which the author directs have had their geospatial analysis published in Tearline.mil.

12 https://en.wikipedia.org/wiki/Intercontinental_ballistic_missile, viewed 18 May 2021.

13 www.atomicheritage.org/history/tsar-bomba, viewed 14 November 2021.

14 www.brookings.edu/blog/up-front/2014/02/27/castle-bravo-the-largest-u-s-nuclear-explosion/, viewed 19 June 2021.

15 www.ourdocuments.gov/doc.php?flash=false&doc= Partial Nuclear Test Ban Treaty, viewed 15 November 2021.

16 *The Corona Story* (Chantilly, VA. CSNR, 2014), chart p. 84; Critical to US Security: The Gambit and Hexagon Satellite Reconnaissance Systems Compilation (Chantilly, VA. CSNR, 2012), p. 16; Gambit, p. 376. Hexagon.

17 Planet website and KH-9 Hexagon Wikipedia article. A Planet Dove weights 5.8 kg. A single KH-9 Hexagon weighted 13,300 kg or as much as 2,293 doves or about 10 times the weight of the entire Planet constellation.   https://en.wikipedia.org/wiki/Planet_Labs#:~:text= Planet%27s%20Dove%20satellites%20are%20CubeSats%20t hat%20weigh%204,ft%29%20and%20envisaged%20environmen tal%2C%20humanitarian%2C%20and%20business%20appli cations.

18 www.nanosats.eu/ Date of Information 4 April 2021, viewed 23 June 2021.

19 A precursor to this trend is the private sector funding of a Finnish SAR constellation to support Ukraine in August 2022.www.iceye. com/press/press-releases/iceye-signs-contract-to-provide-government-of-ukraine-with-access-to-its-sar-satellite-constellation.

20 www.defenseone.com/technology/2017/01/drones-isis/134542/, viewed 23 June 2021.

21 www.youtube.com/watch?v=3yNvI5vJ0Y0, 60 Minutes broadcast 18 September 2022. Viewed 2 March 2023.

22 www.spymesat.com/, viewed 19 May 2021.

23 Podvig, Pavel. "History and the Current Status of the Russian Early-Warning System," *Science and Global Security*, 10: 21–60, 2002.

24 www.nanosats.eu/, viewed 18 May 2021.

25 https://news.satnews.com/2020/11/16/space-flight-laboratory-to-build-three-smallsats-for-ghgsat/ , viewed 19 June 2021.

26 https://globalfishingwatch.org/data-blog/circling-above-point-reyes/, viewed 18 May 2021.

27 The AlexNet and GPT-3 data are taken from Azeem Azhar's *The Exponential Age: How Accelerating technology is Transforming Business, Politics, and Society.* (New York, NY: Diversion Books, 2021), pp. 19–21. The Perceptron data is taken from http://csis.pace.edu/~ctapp ert/srd2011/rosenblatt-contributions.htm, viewed 10 January 2022.

28 www.npr.org/2019/08/30/755994591/president-trump-tweets-sensit ive-surveillance-image-of-iran, viewed 10 January 2021. As the President pointed out, he is the ultimate classification authority in the United States government so his action was legal.

29 Gleeson-White, Jane. *Double-Entry: How The Merchants of Venice Created Modern Finance* (New York, NY: W.W. Norton, 2011). The father of modern accounting is considered to be Luca Pacioli, the first to write about double-entry bookkeeping. He also tutored the young Leonardo da Vinci in mathematics in Milan before Leonardo went to work for Caesare Borgia.

30 https://bigthink.com/surprising-science/radar-moon-astronomy?rebe lltitem=3#rebelltitem3, viewed 19 June 2021.

31 ed. Outzen, James. *Trailblazer 1964: The QUILL Experimental Radar Imagery Satellite Compendium* (Chantilly, VA: Center for the Study of National Reconnaissance, 2012), pp. 40, 44.

32 www.npr.org/2019/12/16/788597818/how-china-is-using-facial-reco gnition-technology 16 December 2019, viewed 17 December 2021. See also Metz, *Genius Makers*, Chapter 15, "Bigotry," pp. 228–238.

33 Matthew Zook, PhD, University of Kentucky, Mark Graham, PhD, University of Oxford, Taylor Shelton, BA, University of Kentucky, Sean Gorman, PhD, FortiusOne. "Volunteered Geographic Information and Crowdsourcing Disaster Relief: A Case Study of the Haitian Earthquake," in *World Medical & Health Policy*, Vol. 2, No. 2, Article 2 (2010). www.psocommons.org/wmhp, viewed 21 May 2021.

BIBLIOGRAPHY

Azhar, Azeem. *The Exponential Age: How Accelerating Technology Is Transforming Business, Politics, and.Society.* (Ashland, Oregon: Blackstone Publishing, 2021).

https://bigthink.com/surprising-science/radar-moon-astronomy?rebellti tem=3#rebellthem3.

www.brookings.org/blog/up-front/2014/02/27/castle-bravo-the-largest-u-s-nuclear-explosion/.

http://csis.pace.edu/~ctappert/srd2011/rosenblatt-contributions.htm.

CIA. *The Intelligence Map Program on Communist Areas 1960–1969*. (Approved for Release: 2018/02/27, C01421958).

www.defenseone.com/technology/2017/01/drones-ISIS/134542.

Eisenstein, Elizabeth L. *The Printing Revolution in Early Modern Europe* (New Haven, CT: Yale University Press, 1976).

Gleeson-White, Jane. *Double-Entry: How The Merchants of Venice Created Modern Finance* (NY: W.W. Norton, 2011).

https://globalfishingwatch.org/data-blog-/circling-above-point-reyes/.

Haydn, Hiram. *The Counter-Renaissance* (New York, NY: Scribners, 1950).

www.iceye.com/press/press-releases/iceye-signs-contract-to-provide-gov ernment-of-ukraine-with-access-to-its-sar-satellite-constellation.

Klein, Gary. *Seeing What Others Don't: The Remarkable Ways We Gain Insights* (New York, NY: PublicAffairs, 2013).

Lanham, Richard A. *The Motives of Eloquence: Literary Rhetoric in the Renaissance*. (New Haven and London: Yale University Press, 1976).

———*The Economics of Attention: Style and Substance in the Age of Information*. (Chicago, IL: University of Chicago Press, 2006).

www.nanosats.eu.

NGA. National System for Geospatial Intelligence. *Geospatial Intelligence Basic Doctrine*. Publication 1.0, April 2018. Approved for Public Release.

www.npr.org/2019/08/30/755994591/president-trump-tweets-sensitive-surveillance-image-of-iran viewed 10 January 2021.

NRO. Center for the Study of National Reconnaissance. *The Corona Story* www.nro.gov/Portals/65/documents/history/csnr/corona/ The%20CORONA%20Story.pdf?ver=BgSn5nPYz45EZ9O_ZF5 7Ow%3d%3d.

———Center for the Study of National Reconnaissance. *Critical to US Security: The Gambit and Hexagon Satellite Reconnaissance Systems Compilation* (Chantilly, VA. CSNR, 2012).

———ed. Outzen, James. *Trailblazer 1964: The QUILL Experimental Radar Imagery Satellite Compendium* (Chantilly, VA: Center for the Study of National Reconnaissance, 2012).

North American Treaty Organization, https://nso.nato.int/natoterm/Web. mvc Definition of Geospatial Intelligence.

O'Connor, Jack. *NPIC: Seeing the Secrets, Growing the Leaders* (Alexandria, VA: Acumensa Press, 2015).

Podvig, Pavel. "History and the Current Status of the Russian Early-Warning System," *Science and Global Security*, 10: 21–60, 2002

Simon, Herbert A. "Designing Organizations for an Information-Rich World," in *Computers, Communication, and the Public Interest* (Baltimore, MD: Johns Hopkins University Press, 1971).

www.space.com?iran-coronavirus_graves_satellite_imagers.html.

www.spymesat.com.

www.tearline.mil/.

https://towardsdatascience.com/alexnet-the-architecture-that-challenged-cnns-e406d5297951#:~:text=AlexNet%20won%20the%202012%20ImageNet,labels%20on%20eight%20ImageNet%20images.

www.ursaspace.com/blog/an_inside_look_at_sar_based_measurements.

https://en.wikipedia.org/wiki/intercontinental-ballistic-missile.

https://en.wikipedia.org/wiki/Partial_nuclear_test_ban_treaty.

https://en.wikipedia.org/wiki/Planet_Labs#:~.text=Planet%27s%20Dove%20satellites%20are%20cubesats%20that%20weigh%204,ft%29%20and%20envisiged%20environmental%2C%20humanitarian%20and%20business%20applications.

https://en.wikipedia.com/wiki/tsar_bomba.

Wolfberg, Adrian. *In Pursuit of Insight: The Everyday Work of Intelligence Analysts Who Solve Real World Novel Problems*. Research Monograph (Washington, D.C.: National Intelligence University, Ann Caracristi Institute for Intelligence Research, July 2022).

www.youtube.com/watch?v=3yNvI5vJ0Y0, 60 Minutes broadcast, 18 September 2022. Viewed 2 March 2023.

Zook, Matthew, Mark Graham; Taylor Shelton, and Sean Gorman. "Volunteered Geographic Information and Crowdsourcing Disaster Relief: A Case Study of the Haitian Earthquake," *World Medical and Health Policy*, Vol. 2, No. 2, Article 2.

Index

Note: Page numbers in *italics* refer to figures.